HUMAN RELIABILITY ASSESSMENT

THEORY AND PRACTICE

HUMAN RELIABILITY ASSESSMENT
THEORY AND PRACTICE

ANTHONY J. SPURGIN

CRC Press
Taylor & Francis Group
Boca Raton   London   New York

CRC Press is an imprint of the
Taylor & Francis Group, an **informa** business

CRC Press
Taylor & Francis Group
6000 Broken Sound Parkway NW, Suite 300
Boca Raton, FL 33487-2742

First issued in paperback 2017

© 2010 by Taylor and Francis Group, LLC
CRC Press is an imprint of Taylor & Francis Group, an Informa business

No claim to original U.S. Government works

ISBN 13: 978-1-138-11618-4 (pbk)
ISBN 13: 978-1-4200-6851-1 (hbk)

Library of Congress Cataloging-in-Publication Data

Spurgin, Anthony J.
  Human reliability assessment : theory and practice / Anthony J. Spurgin.
    p. cm.
  Includes bibliographical references and index.
  ISBN 978-1-4200-6851-1 (hardcover : alk. paper)
  1. Human engineering. 2. Reliability (Engineering) 3. Risk assessment. I. Title.

TA166.S685 2010
620'.00452--dc22   2009027562

Visit the Taylor & Francis Web site at
http://www.taylorandfrancis.com

and the CRC Press Web site at
http://www.crcpress.com

The Blind Men and the Elephant

It was six men of Indostan
To learning much inclined,
Who went to see the Elephant
(Though all of them were blind),
That each by observation
Might satisfy his mind.
The *First* approached the Elephant,
And happening to fall
Against his broad and sturdy side,
At once began to bawl:
"God bless me! but the Elephant
Is very like a WALL!"
The *Second*, feeling of the tusk,
Cried, "Ho, what have we here,
So very round and smooth and sharp?
To me 'tis mighty clear
This wonder of an Elephant
Is very like a SPEAR!"
The *Third* approached the animal,
And happening to take
The squirming trunk within his hands,
Thus boldly up and spake:
"I see," quoth he, "the Elephant
Is very like a SNAKE!"
The *Fourth* reached out an eager hand,
And felt about the knee
"What most this wondrous beast is like
Is mighty plain," quoth he:
"'Tis clear enough the Elephant
Is very like a TREE!"
The *Fifth*, who chanced to touch the ear,
Said: "E'en the blindest man
Can tell what this resembles most;
Deny the fact who can,
This marvel of an Elephant
Is very like a FAN!"

The *Sixth* no sooner had begun
About the beast to grope,
Than seizing on the swinging tail
That fell within his scope,
"I see," quoth he, "the Elephant
Is very like a ROPE!"
And so these men of Indostan
Disputed loud and long,
Each in his own opinion
Exceeding stiff and strong,
Though each was partly in the right,
And all were in the wrong!

John Godfrey Saxe (1816–1887)

Contents

Preface

THOUGHTS OF AN ENGINEER

I do not know if a preface ought to be written by someone else, maybe it should? However, these are some thoughts of someone who has been connected with engineering. I love engineering. It would be nice to be like Leonardo, an engineer, an artist, a conceptualist. My life has been interesting—as a child in India, a schoolboy here, there, and everywhere, and at university in London. The sport of my life was rugby, a fantastic team sport. I played it in England and the United States. But you ask, what can sport offer? So many things: support for your fellow players and appreciation for your opponents, and thanks to the referees for their involvement, giving their time to control a bunch of roughneck gentlemen. Remember, rugby was once called a "hooligan's game played by gentlemen!"

I guess that most engineers consider that they can do anything. What an egotistical view! But maybe it is true; after all, Leonardo was an engineer. Like most people, my views are colored by my experiences. I was taught engineering by a group of capable lecturers, some better than others, but all good. The same goes for my teachers at school. I think that in some ways I was putty in their hands. However, the responsibility for the materials within the book is mine.

It is not clear to me what kind of training one should have to be involved in the topic of human reliability. The topic appears to be the close companion of equipment reliability, because the overall reliability of a power plant, system, aircraft, factory, ship, and so forth, cannot be limited to just the reliability of pieces of hardware. The human is involved in all aspects of work. Some parts of the relationship between men and machine have been considered, and to some extent codified, as human factors. Some other aspects have been considered in that we train people to operate machines, and to support them in their endeavors we supply them with procedures. Then to compensate for the possibility that human error may creep in, we developed the concept of the team with a leader to ensure that the most important things are tackled first.

The real heart of human reliability is to find credible ways of helping designers, management, operators, and authorities to be able to help increase the safety and profitability of technological systems. Human reliability coupled with probabilistic risk/safety assessment introduces people to a thought process to perceive risks in operation and help define ways in which the risk can be reduced. Also, because failures in an operating situation can lead to loss of output as well an unsafe state, this directly affects the operating cost of running the equipment. So, the idea of separating safety from availability is not logical. The object of probabilistic risk assessment (PRA)/human reliability assessment (HRA) is to give management the tool to assess the risk factors associated with operations. Having identified the risks and their probability, then one can take the next step to decide what, how, and when changes should be made. The lesson learned by the U.S. utility and U.S. Nuclear

Regulatory Commission (USNRC) from the Three Mile Island (TMI) Unit 2 accident was that the risks were higher than they thought, that PRA was a useful tool for defining operating risk, and that the environment under which the humans operated needed to be taken more seriously. In fact, many of the recommendations made by the Kemeny President's Commission on the accident at TMI have been implemented. The commission's report (dated October 30, 1979) was followed by a whole series of reports and recommendations. Most were directed toward improving the operating environment: control room layout, procedures, training, and the use of nuclear power plant (NPP) simulators for exposure to multiple failure scenarios. More will said about this later in the book.

Prior to the TMI Unit 2 accident, the first comprehensive PRA study (WASH 1400) was released in 1975. It is interesting to note that it was claimed that the TMI accident sequence was present in the set of accident sequences. I do not believe that the message contained within WASH 1400 was as decisive in the area of human performance limitations as was the set of recommendations following the TMI accident. One could argue that the methods HRA used in the study, the way the HRA expert was used, and the lack of experience in the NPP field of HRA/human performance effects underestimated the impact of the human reliability on the core damage probability estimates. One reason for this shortcoming could be aimed at the failure of the industry at the time to realize the importance of humans relative to safety. Engineers have a tendency to shy away from understanding human responses and are inclined to design around the fact that humans are needed. There was an assumption that the automatic safety systems were adequate and that reliance on operators was not required. Luckily, the emphasis on human performance has grown. The USNRC has a comprehensive program directed toward understanding human performance (O'Hara et al., 2008). This is in contrast to the early days, when it appeared that the only safety requirements as far as control rooms were that each safety system had to have separate indicators to show its status.

I was for many years a designer of control and protection systems for various power plants from conventional power stations to nuclear. I must admit that I did not totally recognize the special needs and requirements for operators. In the process of designing the control systems, we considered an accident similar to that which occurred at TMI, but many years before TMI. As result of considering pressurized water reactor (PWR) dynamics, it was decided that it was not necessary to include extra instrumentation to detect the onset of conditions that would lead to core damage. We thought that the information available to the crew would be sufficient to predict the situation and take the necessary actions. The knowledge and understanding of the analyst and the operator are not the same. We were wrong and should have tried to understand the operators' views of the developing situation. It is interesting to note that Dekker (2005) makes the same kind observation and frames the issue as the "view from the cockpit is not the same as the accident investigator." He was talking about aircraft accident investigations, but the point is well made and taken. In performing an HRA study, one needs to understand and model this "view from the cockpit!"

Acknowledgments

Over the years, I have worked with a number of human reliability assessment (HRA) experts and others, while carrying out studies in the nuclear and aerospace fields as well as trying to develop Institute of Electrical and Electronics Engineers (IEEE) standards. The interactions have been very useful. I cannot say that there was perfect agreement by all on my ideas or equally the acceptance of the ideas of all of others. I would like to acknowledge a few and apologize to the others whom I have not acknowledged, because the list would be longer than the book. Some operate in the United States, and others are outside of the United States. Thanks to Attila Bareith, Erik Hollnagel, Bill Hannaman, Vojin Joksimovich, Mike Frank, Paul Amico, Doug Orvis, Parviz Moeini, Gueorgui Petkov, Cathy Gaddy, Elod Hollo, Arthur Beare, Pierre LeBot, Bengt Lydell, Mary Presley, Gareth Parry, and David Worledge.

Thanks also to my wife for her patience and help, especially during the time I was writing the book, and to my family, including my sometime colleague, Jonathan Spurgin. Thanks to my editor Cindy Carelli, production coordinator Amber Donley, and project editor Amy Rodriquez at Taylor & Francis for their enthusiasm and support.

I also must thank Google and Wikipedia for access to the world of Internet research.

The Author

Anthony Spurgin is an engineer whose main interest is in human reliability assessment and its application to various industries to improve safety and economical operation. Educated in England, he received his BSc in aeronautical engineering from the University of London. After university, Spurgin was a graduate apprentice at an airplane factory and later performed aeroelastic calculations for a number of different aircraft. Subsequently, he moved to the nuclear power generation industry, where he has been involved in the field of human reliability assessment, from the point of view of the development of methods and techniques and also in applying those techniques to the enhancement of the safety of nuclear power plants (NPPs) in a number of different countries. A central part of his experience is related to the design of control and protection systems for NPPs. He has been the principal designer of control systems for pressurized water reactors (three-loop plants) and advanced, high-temperature, and Magnox gas-cooled reactors. He has also been involved in the study of conventional fossil power plants, including oil and coal-fired plants. He was also involved in the design of various test rigs, covering steam generator behavior and reactor loss of flow (LOCA) experiments. His time at reactor sites was spent redesigning systems to get them to operate successfully.

1 Introduction

1.1 PURPOSE OF THE BOOK

The purpose of this book is to cover the topic of human reliability assessment (HRA) in some depth from the viewpoint of a practitioner. There have been a lot of changes in the topic over the last number of years. Because HRA is still developing, it has elements of controversy in terms of the appropriate methods for the representation of human failure probability. Even the term *human error* causes problems for certain individuals. The idea that human error is a random event is not acceptable, and the concept that humans can be set up to fail due to the context or situation under which they are operating is gaining credibility. This concept means that one can do something to avoid human errors, and it is not a case of firing "error-prone" individuals. In the study of human error, simulators are now seen as a valuable resource, and their use in this field has gone through a change in attitude. This book will cover aspects associated with data and data sources, choice of methods, training of individuals, use of simulators for HRA purposes, and relationships between psychology, human factors, accident analyses, and human reliability.

In writing a book about this topic, what to include and what to leave out is an issue. Some of the topics covered here in a few paragraphs are covered by many books. For example, there are many models attempting to cover only the prediction of human error probability. In one research document, Lyons et al. (2005) documented some 35 methods and techniques, and I would guess that this number is an underestimate. I do not intend to cover each method and technique. A number of HRA methods have been selected to discuss which are currently in use in the nuclear industry in the United States and elsewhere. Many of the approaches discussed here lean heavily on nuclear energy applications, but they are adaptable for use in other industries. The nuclear energy field is the wellspring for the development of ideas in both probabilistic risk assessment (PRA) and HRA.

The idea of human reliability has grown out of the development of equipment reliability as a science and the evaluation of systems. The concept of risk has played a major role in its development. Once one starts to consider the risk of an accident to an operating plant leading to both economic loss and loss of life, eventually one is forced to consider the role of organizations, decisions made by both managers and staff, and, of course, a risk–benefit relationship of design and operational decisions.

At one time, the failure of equipment dominated the field of reliability; in other words, equipment failed because components within a system failed. However, as systems became more complex, the role of the human changed from being just the user of a piece of equipment to being more directly involved in its operation. An example of a simple piece of equipment would be a plow, and the failure of a wheel or the plow blade could be seen as an equipment failure. In earlier days, no one paid attention to the way the blade material was chosen or how the wheel was built.

Therefore it was an equipment failure, not a human failure. Later, the concept of a system was developed with the combination of both man and equipment being involved in the design and operation of the system. So, for example, a power plant is seen to be a system. Initially, the same bias was present in that the power plant was dominated by consideration of the components, such as pumps and valves. A theory of component reliability evolved in which there was an early failure, a random failure, and final failure probability, the "bathtub failure model." The science of equipment and component failure grew, but there was little attention paid to the impact of humans. Over time, the role and importance of human contributions has grown; for the early PRAs, the contribution of humans was set at about 15%. Now the contribution has grown to 60% to 80%. The reason is a growing understanding of human contributions, but also equipment reliability has increased due to better choices of materials and attention given to maintenance. Even here one is aware of the human contribution. The growth of interest in HRA and its methods and applications is the reason for this book. One might ask, why does one consider HRA? One considers HRA within the framework of a PRA to assure the regulatory authority, the public, and the plant owners that the plant is safe to operate and the risk is acceptable according to the rules and regulations of the country. In the case of the nuclear industry, it was and is important to demonstrate that nuclear power plants are safe to operate.

The use of PRA/HRA is expanding to more directly include management in the process, because the safe and economic operation depends heavily upon the decisions of management. Initially, a PRA/HRA was produced and used to meet regulatory requirements, and afterward the documents were stored and the potential uses of the study were neglected. Many PRAs collected dust on the shelves of libraries.

Of late, the PRA results have been transformed into a much more usable form so that one could examine the consequence of change in the plant risk following a configuration change. These results could inform the management and the regulator of the change in risk as a result of changes. If investigations were undertaken ahead of a change, management could make judgments as to the duration of the change to see if this was tolerable or not. If it is not tolerable what should one do? The number of alternatives, such as reduce power or shut down, increase surveillance of the plant, or even bring in standby equipment could be studied for the purpose of reducing the risk to a tolerable level. Once the assessment is made, the "best" choice could be selected.

One thinks of the power plants in the following manner: They are designed and built, staff is engaged and trained, and then the plant is operated for many years and nothing changes. In practice, changes are going on all the time: Staff members leave and are replaced, equipment fails and is replaced, digital replaces analog equipment, the plant is upgraded, equipment ages, steam generators and reactor vessel heads are replaced, new control drives are introduced, and so forth. In addition, management wants to produce more power consistently within the operating license by proving to the licensing authority that the power output can be increased to a new level by changes to turbine blading and nozzle design and can be done safely. All of these things involve changes to the operating envelope and therefore affect the operating personnel. Steps have to be taken to upgrade personnel performance by training, enhancing procedures, and making changes to display and control equipment, including operator aids.

All of the above means that the HRA techniques have to improve to match the needs of management and to prove to the regulator that the changes are acceptable.

Much of the above applies to other industries, where change is going on and the concerns of the public are matched by the actions of the appropriate regulators. The drive is always an increase in the effectiveness of the plant and equipment (more output for the same or lower cost) balanced against the concerns of the public seen through the activities of the regulators. At the center of all of this is the human, so one must consider the impact of the decisions and actions made by management, designers, and constructors on the operators. This book links many of these things together in the studies of accidents in Chapter 8. HRA is still developing, and various groups are thinking about developing different tools and how to help the industry move forward to satisfy its economic and safety requirements. We are like the blind men from Indostan; we each see part of the whole but are not quite seeing the whole HRA picture in the development process. This, in a way, explains in part why there are so many HRA models and why people are adding to them each day.

1.2   CONTENTS OF THE BOOK

The book consists of 14 chapters covering a variety of topics. The first four chapters provide an introduction and background, and then cover risk concepts and HRA principles. Chapter 5 covers a range of current HRA models. Over the years there have been a number of proposed HRA models and methods, some of which have disappeared for one reason or another. Even the current models may not continue to be used in the future, or their use may grow. Often, which models continue to exist depends on the support of organizations, such as regulators, and not necessarily on the intrinsic qualities of a model or method.

The next two chapters (Chapters 6 and 7) deal with HRA tools and provide a critique of the set of models identified in Chapter 5. As a user, one needs to know not only about a specific model or method, but what are the strengths and limitations of the model. A particular model may have been chosen because of the organization to which one is attached, or perhaps because of the same regulatory influence, but the analyst ought to understand the limits of this decision and perhaps modify the results to be more reflective of his or her application. One of the objectives of this book is provide some ideas and concepts to broaden the decision base of working HRA specialists. The book can also provide the decision makers with the research either to back up their decisions or lead them to change them.

Chapter 8 covers a number of typical accidents. The accidents have been selected from several different industries, from nuclear to aerospace to chemical to railways. There are many commonalities about accidents, and the purpose here is to pull some of these together so that the reader can understand the circumstances and forces at play that led to the accidents. This understanding can enable one to better analyze situations in technologically risky fields with the objective to reduce the number and severity of accidents. A review of this chapter will bring home the important role of management in the safety of installations.

Chapter 9 considers each accident covered in Chapter 8 from a prospective point of view. This is somewhat difficult, because the accident contributors are known, and this is a significant influence on the process—but the attempt is to show how HRA procedures might be used and then see if the result is in any way close to actuality. The important part is to look at the process in light of what needs to be examined. One problem found during this prospective view is—and this may be one that would exist even in practice—the availability of access to all information on the design and operation of the plant or equipment. Clearly, the accident reports are limited as to what is published, so not all of the information, like interviews of plant personnel, may be available. Also, the needs of the accident investigators may not be the same as the PRA/HRA analysts; in fact, this is often the case. One accident studied was the accident at a Bhopal chemical plant. In this particular case, even the accident investigators did not have access to plant personnel until a year after the accident. The Indian government was very controlling in terms of the accident details, data, and interviews. Why was this so? Your guess is probably as good as mine, but it was difficult for investigators at the time to really understand the accident causalities. It seems, however, possible to draw some useful conclusions from the accident, even if the details of the accident are muddy.

I have taken the opportunity to illustrate the steps in an HRA study related to the International Space Station (ISS). Chapter 10 is based upon work undertaken on an HRA study for the ISS and involves the application of the holistic decision tree (HDT) approach along with the steps taken to incorporate NASA knowledge and experience into the study. NASA-based knowledge and experience was upon flights to and from the ISS, issues that had occurred to the ISS, and knowledge stemming from simulator sessions with interactions between astronauts, flight controllers, and instructors. It was this knowledge and experience that outside reviewers thought was necessary to carry out a useful HRA study. One would have to agree with them, and it reinforces the idea that the best HRA studies rely on the knowledge and experience of the domain experts rather than just the knowledge experts.

One of the key elements in any HRA study is the availability and use of data. Chapter 11 deals with data sources and data banks. Often the analysts are capable of sorting out the sequence of events that can influence a crew in their performance responding to an accident, but their selection of an HRA model either leads to a built-in data bank of questionable applicability or to not knowing where to obtain data to complete the analysis. So the question comes up, is there a useful database for a given application or does one have to use expert judgment estimates? Chapter 11 tries to answer some of the questions raised by users.

As part of the process of training improvements following the lessons learned by the utility industry after the Three Mile Island Unit 2 accident, the U.S. Nuclear Regulatory Commission (USNRC) required each nuclear power plant to have a full-scope simulator. Crews are trained in various ways using simulators, and data and information derived from the responses of crews to accidents are not only useful for the preparation of the crews in case there is an accident, but also can be used for HRA purposes. Chapter 12 discusses simulators and data collection processes.

Simulator scenarios designed for both requalification training and for HRA purposes can be used for direct data circumstances, where feasible, or for informing experts to help in forming judgments related to estimating human error probabilities. Electricité de France (EDF) has been collecting simulator data and information, and using expert judgment for a long time. Currently EDF is using this data and information to obtain human error probability (HEP) data for use with their latest HRA method called Methode d'Evaluation de la Realisation des Missions Operateur pour la Surete (MERMOS). Some details of the MERMOS method are discussed in Chapter 5 along with other HRA methods and models.

A critical review of the current use of simulators for training is presented in Chapter 13. The author thinks that simulator usage could and should be improved to enhance safety and at the same time provide useful information for the management of power plants as far as quantifiable information on the performance of the crews is concerned. Other industries use simulators, such as the aircraft industry. Simulator usage in the aircraft world could be increased to compensate for the shift to more automated aircraft. Pilots could spend more time responding to accidents when the automated equipment malfunctions during takeoff or landing or during weather-induced problems. Nuclear power plants are quite compact and have simulators for training that can be fairly easily designed. The same is true for fossil and hydro plants. However, many chemical plants involve distributed equipment and different processes, so the same kind of simulator is more difficult to design because of the number of distributed human–system interfaces. However, generalized part task simulators could be used for individual chemical plant units provided suitable generators to model input disturbances from other units are included in the simulation.

Chapter 14 contains discussions on the current state of HRA and conclusions on what research might be carried out on the predictability of human errors, and how context effects on human error can be estimated and to what degree. A basic issue with estimation of HEP is that it still depends greatly on expert judgment. The use of simulators for gathering data and insights can be useful in the HRA process; some suggestions are made in this chapter for the use of simulators to extend the knowledge of HRA effects resulting from, for example, human–machine interface (HMI) design. It is hoped that by the time the reader gets to this point in the document, the understanding of the HRA elephant will be a little clearer.

Literature references follow the final chapter, and the Appendix provides database references.

1.3 POTENTIAL USERS OF THE BOOK

Users of this book could come from industries that are exposed to the risk of causing injuries and death. Personnel within the industries that have already had major accidents are candidates. It is suggested that members of management teams consider reading portions of the book, such as Chapter 8 on accidents, because they ultimately make the decisions about the design, construction, and operation, including maintenance, of their plants. The decision-making process may be implicit or explicit, but nonetheless, the buck stops with them.

Students interested in the field of human reliability and the consequences of human error should read this book, either those at university level, or just students of life. Nuclear energy is going through a resurgence, and engineering students might be thinking of joining a nuclear group of one kind or another. It could be useful for those students hoping to gain some insight into this topic. There is a need for additional research into this topic, so MS and PhD students could well find many topics in the HRA field, and this book could be a useful aid in this endeavor.

1.4 AUTHOR'S PHILOSOPHY

One should have a basic philosophy guiding one through the machinations of life. In dealing with human reliability, one should have a concept of what one believes are the driving forces influencing the determination of failures called human error.

I believe the overwhelming human drive is to do well, and that it is the responsibility of management and organizations to match that drive and help individuals achieve that objective on a continuous basis. This whole approach determines my approach to the science and the art of human reliability assessment (HRA). I do not believe humans set out to make errors, but rather errors occur due to the context into which they are placed. There appear to be a number of books on the market that cover this approach. For example, Dekker (2005) talked about the view from the cockpit in order to understand why pilots get into accidents. Very few pilots wish to kill themselves or their passengers. Train drivers also do not wish to commit suicide. Clearly, there are people who commit suicide and occasionally one happens to be a pilot; one Egyptian pilot reportedly did just that leaving Kennedy Airport in New York, but this is the exception to the rule. In the case of Bhopal (see Chapter 8), some of the investigators believe that the accident was caused by a disgruntled employee, but perhaps even this accident was avoidable if this employee's condition was carefully considered earlier by management.

In judging how to prevent accidents, one needs to consider the position of the operator and the demands made upon him or her. The single-track railway is a perfect example of a setup for a human error and the dire circumstances that can follow an operator error. The method of running a single-track railway is to have trains running in both directions, with a limited number double tracks to allow trains to pass. This arrangement is driven by economics—it costs too much to have two tracks all the way. Depending on the safety arrangements, signaling duplication, and controls, safety does depend on train drivers not making mistakes; however, to avoid accidents, the overall reliability of the rail operation needs to be considered. To improve the reliability of the railway system, one needs to combine both human and equipment reliability aspects acting together. Actually, it is amazing that train drivers are not involved in more accidents, given the lack of design and operational considerations by railway management. The nuclear industry considered a single failure to be acceptable design criterion, including human failures, whereas the railway business seems to depend on drivers not making errors. In a recent Los Angeles case, the driver was thought to be distracted by text messaging on a cell phone, which led to a freight train and a passenger train colliding, killing and injuring several people including the passenger train driver. In my opinion, the system for train protection was inadequate, and the operator ought to have been actively as well as passively warned by the use of a red light. Interestingly, the old-fashioned method of stopping the train with a mechanical lever actuated by the signal that disconnected the power to the train drive would have accomplished the task quite well.

Systems need to be designed to work with humans to support them in meeting the goal of operating systems safely and economically. Not only do they need to be designed, they need to be operated to support the same ideal. Management needs to be involved to ensure that operators are assisted, rather than being blamed for situations that go wrong. It is not possible to see into the future and prevent all accidents, but management should be involved in monitoring how things are working and whether there is any deterioration in the operation of the plant. Often when no accidents have occurred, there is the idea that they cannot occur, and therefore investment in safety controls and maintenance of equipment can be reduced. These are the actions (or, rather, lack thereof) that can precede an accident. The set of accidents described in Chapter 8 identify clearly the role of management in accident initiation and progression.

2 Background

INTRODUCTION

Historically, the first serious consideration of human reliability and risk may have been made during the Cold War (1945 to 1989). The United States was monitoring the possible launch of Russian missiles against the United States and the North Atlantic Treaty Organization (NATO). Radar screens at places like Filingdales, Yorkshire, United Kingdom, were being inspected on a continuous basis to see if rocket launches were occurring, and when they did, they tracked their trajectories. The reliability aspect was to distinguish very quickly between an aggressive launch, which necessitated a response, and some activity that was quite harmless. Another aspect was what to do about a single launch that could be spurious, so the operators had to act quickly and be very correct in their interpretations, especially as far as England was concerned, because the flight time was about 5 minutes. Associated with human reliability is the element of risk. In this particular case, the consequences of being wrong could lead to the annihilation of cities and nuclear war.

2.1 EARLY DEVELOPMENTS

Another human reliability event, similar to the above, was the assembly of nuclear weapons to ensure that they worked when required and not inadvertently at some other time. Again, the factor of risk was associated with human activities. The human reliability method developed by Swain and Guttman (1983), Rook (1962), and others was later published under the sponsorship of the U.S. Nuclear Regulatory Commission (USNRC) as the "Technique for Human Error Rate Prediction" (THERP) in draft form in 1979. The atomic bomb, in one form, is a spherical shell made of uranium-235 (U235), which is surrounded by a number of highly explosive charges, the purpose of which is to implode the shell into a compact sphere in a fantastically short time. The shell is not nuclear active, but the compact sphere becomes critical with a high doubling rate, and this is what leads to an explosion. Clearly, the timing of the implosion charges is important, but also of importance is the series of tasks to set up the locations, connect the related electrical connections, and so forth. Any failure to perform the tasks could lead to a premature explosion, which may not be large but could radiate the persons assembling the bomb.

The next significant step in the expansion of human reliability considerations was the formation of a team to study the safety of nuclear power stations under the auspices of the U.S. Atomic Energy Commission (AEC), later to become the USNRC. The design of nuclear power plants prior to this time was carried out using what was called "design-basis" accidents coupled with a failure concept known as the "single-failure" concept. The study group's leader was Norman Rasmussen (Massachusetts Institute of Technology), with the day-to-day leader being Saul Levine. In addition, a number of current luminaries were involved in the study. The study was the first probabilistic risk assessment (PRA), and the report was released in 1975 as WASH 1400. One should also mention the Accident Initiation and Progression Analysis (AIPA) study (Fleming et al., 1975). This is the first PRA study, considering a high-temperature gas-cooled reactor (HTGR), as opposed to WASH 1400 that covered light-water reactors (pressurized and boiling water reactors). Personnel from both studies have influenced the development of PRAs. Vojin Joksimovich was the director of the AIPA study.

As far as human reliability was concerned, the WASH 1400 was based upon the THERP method developed by Alan Swain, plus the use of expert judgment. The AIPA study was based on a time-dependent success model derived from reactor control experiments involving manual control in the event of control equipment failures. General Atomics carried out studies on a hybrid simulation (using analog and digital computers) of an HTGR plant for the purpose of designing plant controls. G.W. Hannaman converted the success model developed into a failure model more suited to the AIPA study. As far as is known, this was the first use of a time-based model of operator actions, or time reliability curve (TRC).

The elements of the AIPA HRA were based on a time-response model seen during the control studies (i.e., a time delay in which there was no response by the operator to changes to displays occurring as a result of a change in the plant). This was then followed by an exponential curve representing success up to a limit, which was less than 100%. This latter limit was to account for the fact that under some circumstances, the operator or crew could fail. Although this model does not reflect the cognitive processing going on in the mind of the operator or any of the environmental or context influences, it indicates a typical response of operators to dynamically changing situations. Many later simulator studies have borne out this type of response. However, where does this take us? Unfortunately, not very far, because at the time it was not possible to calibrate the curve, and we could not generate predictions of possible future operator actions.

The HRA handbook by Swain and Guttmann was a major piece of work that was based on the concept of task analysis. The handbook accounted for all of the aspects of a complete HRA approach, including dependencies between operators. In the opinion of many, the very completeness of his approach has been a constraint on future developments. Thus many of the later approaches and methods tackled parts of the HRA area considered by their developers not to be adequately covered in the handbook.

Any task can be broken down into a number of subtasks. Many books and papers have been written on the topic. One is by B. Kirwan and L.K. Ainsworth (1992). Task analysis is a human factors tool used to examine the working relationship of operators in performing a task, and traces such things as what number of things are expected of the operators, what tools do they use, and what information is relevant. The THERP approach mirrors the task analysis approach, with each element or subtask being modeled as a reliability element. So the whole task is composed of these elements, which together are used to predict the reliability of the operators undertaking the task. However, because there was more interest in the failure of the operators to perform the task correctly, given that the atomic bombs might accidentally explode or fail to work when required, the negative or failure probabilities were used.

2.2 USNRC-SPONSORED HRA-RELATED STUDIES

Both the industry and USNRC were interested in the role of the operator in taking safety actions versus that of equipment. These were the questions being asked: What is the appropriate role of the operator in responding to accidents? What should be carried out by safety protection systems? The industry developed, under the auspices of the American Nuclear Society, a proposed standard (ANS N-660, 1977) covering this issue, but it needed to be supported by actual results. The USNRC then initiated a program to use simulators to gain information on the success of control room crews in responding to accidents. Boiling water reactor (BWR) and pressurized water reactor (PWR) simulators were used for this study. At this time, it should be remembered, the use of simulators was limited, and there were very few of them. Emergency operating procedures (EOPs) were event based, and the scenarios used during simulator training were very simple, rather than of the multifailure type used later. The program was executed by Oak Ridge National Laboratory (ORNL) and General Physics (GP). Many reports were issued on the results of the studies (Kozinsky et al., 1984). Later, the N-660 document was replaced by ANSI/ANS 58.8, 2008 being the reaffirmed version.

This program had a big influence on HRA development and gave rise to the idea of using the TRC in a number of HRA methods. Time was considered to be a key parameter in determining operator error potential. In 1979 following the TMI accident (March 1979), the Institute of Electrical and Electronics Engineers (IEEE) organized a joint IEEE/NRC 1979 meeting to review Swain's human reliability handbook, among other things. The result was that the meeting was critical of the THERP methodology in that it did not explicitly cover cognitive actions, and it was suggested that the handbook should be modified to address this issue. The TRCs were considered at the time to be measures of cognitive activity. Thus the final version of the handbook addressed the cognitive issue by including a TRC not so much based upon the ORNL/GP results, but drawn by Swain with advice from experts. The ORNL/GP TRCs deal with the idea of nonresponse, and the curve measured how many of the crews were successful, because failures were not included in the TRCs. To account for the possibility of both delayed actions and failures by the crews, Swain extended the TRCs to include low probability values for longer times.

The USNRC also sponsored a number of other HRA studies or reports shortly after the publication of the handbook. One was a report by Hall, Wreathall, and Fragola (1982), which was closely related to TRC concepts. This report covered post-accident decision errors, introduced the idea of operator action trees, and analyzed the ORNL/GP TRCs from the viewpoint of Rasmussen's skills, rules, knowledge (SRK) model (Rasmussen, 1979). Another report, carried out by Embrey and associates, introduced the success likelihood index method (SLIM) (Embrey et al., 1984). This method is really quite like a later-generation HRA in that the emphasis is on context leading to estimation of human error probabilities (HEPs). Context here is defined by a collection of weighted performance shaping factors (PSFs). Expert judgment is used to select the weighting factors. The method was embedded into computer code. The HRA knowledge expert would use domain experts to generate the weighting factors. The domain experts could be operators or instructors. The other feature of SLIM was to use a modifier to further shape the HEP. It is found in psychological investigations that human estimated do not correct yield exact results, but tend to be biased. An adjustment for this is given in the document and the computer code. Another funded project at this time was a report of various expert judgment approaches including direct estimation (Comer et al., 1984). The report also covers how expert judgment solicitation sessions should be set up and suggests the minimum numbers of experts to use to improve the accuracy and repeatability of the results. Expert judgment can suffer from uncertainty, especially if not carried out in a well controlled environment. However, but it is critical to the use of HRA. Without it, it would be difficult to carry out an HRA study. In fact, all HRA methods depend to a greater or lesser extent on expert judgment. More about this is presented in Chapters 5 and 7.

The Three Mile Island (TMI) accident in March 1979 (Kemeny, 1979) was a landmark event in the development of nuclear power in the United States, and its impact rolled over to other countries. The main effect of TMI was to change the whole attitude on the role and importance of humans in the safety of nuclear power plants (NPPs). The reaction to the TMI, Unit 2 accident, was to change the whole view of nuclear power operations with respect to operators, control room displays, simulators, and training. One very important change was to change the EOPs to symptom-based and to increase the importance of EOP usage.

2.3   EPRI-FUNDED HRA PROJECTS

Following the TMI Unit 2 accident, the Electric Power Research Institute (EPRI) became interested in developing HRA methods to support the electric power industry in this post-TMI period. Worledge (the EPRI program manager) started with calling for a systematic way of organizing HRA within the PRA, and this led to the development of the Sysematic Human Action Reliability (SHARP) method (Hannaman and Spurgin, 1984a). Later, versions of this basic method were developed by IEEE (IEEE Standard 1082, 1997) and later updated by EPRI in the form of SHARP II (Wakefield et al., 1992). The next initiative by EPRI was to develop the human cognitive reliability (HCR) correlation (Hannaman et al., 1984b) to cover the cognitive actions thought to be missing in the THERP approach. This method was widely used within the nuclear power industry, including in countries outside of the United States for application to PRAs and probabilistic safety assessments (PSAs).

EPRI also funded a couple of projects around the same time to examine the availability of human reliability databases and review work on HRA and human factors aspects currently available. Worledge was looking for how much had been investigated prior to this time. These reports, like the HCR reports, were draft documents.

It was thought by EPRI that HCR was interesting in that it integrated the concepts of Rasmussen (Rasmussen, 1979) as far as skill-, rule-, and knowledge-based behavior was concerned with a normalized TRC, which could be used for HRA operator actions. By using a normalized TRC, the developers said that the underlying characteristics of operator responses were much the same for all operator actions; the only difference was that it took a shorter or longer time and the actual scaling of the normalized curve was achieved by multiplying the time scale by the mean time for crews to accomplish an action.

It was decided by EPRI that the fundamental assumptions within the HCR formulation needed to be proven. It was on this basis that EPRI started the operator reliability experiments (OREs) program (Spurgin et al., 1990a), and it was supported by six NPP utilities. Both PWR and BWR utilities were in the group. The program would be based on the utilities' simulators, and data would be taken on both operator time responses and observations of their actions. Chapter 12 contains information about simulator data collection projects, including ORE. The results of the ORE project were published; unfortunately, the reports were limited to EPRI utilities (utilities that supported EPRI, i.e., paid dues to EPRI). The results coming from the ORE program showed that not all of the assumptions used to derive the HCR correlation were supported, but some were.

There were a number of spinoffs from the ORE program, including the development of more automated means of collecting data on operator performance that could be used for operator training, insights into what items of the accident scenarios could be influenced and what could not be, and how to design experiments to be carried out at simulators. The ORE observations also led to an understanding of what really determined operator performance during an accident, and that led to the formulation of the Holistic Decision Tree HRA method.

Another project was requested by EPRI on reaching a better understanding of operator recovery actions not guided by EOPs, and how much credit one could attach to these actions. A draft report was presented to EPRI, but it was not published as an EPRI official report. Recoveries reviewed included subjects such as the recovery of diesel generators in the face of failures.

An EPRI report (Spurgin et al., 1989) based upon the ORE report data was one that gave ORE/HCR curves. This report was issued to help utilizers during responses to independent plant evaluations (IPEs), which were called for by the NRC.

General Physics (GP) was a participant in the ORE program, and the company was asked by EPRI to try to use the ORE results to generate an HRA method. Beare (GP) developed the cause-based decision tree (CBDT) method nominally based upon ORE results, but really more correctly based upon a restatement of THERP. A draft version of the report was issued to the members of the ORE team. A final report was by EPRI, TR-100259 (Parry et al., 1992), and when combined with both the ORE/HCR results (EPRI NP-6560 L) together made up the Beare draft report.

This was the last major HRA report issued from EPRI until fairly recently. EPRI's involvement in nuclear power diminished following the U.S. nuclear industry's lack of involvement in new construction. Some work was going on during this quiescent period; for example, there were a number of contracts initiated by utilities to combine their funds with funds from EPRI to get more leverage to accomplish useful projects. There were two projects. The first was with Texas Utilities to develop an HRA calculator tool (Moeini et al., 1994) and the second was the development of a data collection system for simulators sponsored by Pacific Gas and Electricity. There were a number of developments of this project, leading to CREDIT (Spurgin and Spurgin, 1994b).

2.4 OTHER ORGANIZATIONAL INVOLVEMENT IN HRA

Other organizations, notably consulting companies, were involved in the use of HRA methods. Two companies that were heavily involved in PRAs were Pickard, Lowe, and Garrick (PLG) and Science Applications, Inc. (SAI). Later, SAI grew into SAIC (Science Applications International, Co.). PLG used a development of Embrey's SLIM method called FLIM. SAIC used a development of the TRC approach (see the book by Dougherty and Fragola, 1988). Several HRA developers were aware that there was a need to develop HRA methods more in tune with the reality of human responses to accidents.

Outside of the United States, the French and the British were involved in HRA. Electricité de France was involved in carrying out experiments with simulators initially from a human factors viewpoint. Later, they moved to use the experience on early simulator results to generate a modified version of Swain's TRC for their 900 MWe PWR PSA study. EDF also had an arrangement to share experience on simulator sessions with EPRI. In fact, EDF and EPRI agreed to run similar simulator sessions and compare the results on how operators responded to these accidents.

The British made use of an HRA method developed by Jerry Williams based upon his experience in various human factors fields. The database generated by Williams was included in his papers. More about the human error assessment and reduction technique (HEART), which is Williams' method, will be discussed later in this book. EDF research has abandoned the early HRA work and developed a new approach, also based on the simulator observations of control room crews, called Methode d'Evaluation de la Realisation des Missions Operateur pour la Surete (MERMOS). MERMOS is also covered in the later chapters of this book.

A number of other countries have based their HRA studies upon the HRA handbook (Swain and Guttman, 1983), EPRI's HCR correlation or use of simulator results based upon ORE methods of data collection. These studies came some time after the original work was done in the United States.

As mentioned above, a number of researchers realized the limitations of the previous HRA methods and were beginning to consider other methods and models. Dougherty was the first to give voice to this recognized need in a paper entitled, "Human Reliability Analysis—Where Shouldst Thou Turn?" in 1990. He came up with the idea of the second-generation HRA models. The first generation included HRA models such as THERP and HCR. In later chapters, this book covers a number of HRA models and discusses their pros and cons. THERP and its derivatives continue to be a force in the HRA world despite their shortcomings.

3 Plant Safety and Risk

INTRODUCTION

Everyone is concerned about the safety of high-risk technological power plants, refineries, and nuclear power plants (NPPs). Organizations should be studying methods to help prevent accidents. Safe operations and safety in general were concerns a long time before the nuclear industry was established. People are concerned about the safe operation of mines, factory equipment, trains, dams, and road transport. No matter how many precautions one may take, the environment where one lived or worked is subject to risks that are difficult to avoid. There is no wish to go into all of the steps in getting society as a whole to appreciate the presence of risk and the steps that should be taken or are being taken on behalf of society. This would call for a different book.

In this chapter, the development and implementation of safety ideas for nuclear power are considered. Safety ideas and their implementation have changed and vastly improved over the period in which nuclear power has developed. When nuclear power was first developed, it was closely associated in people's minds with the atomic bomb. Power plants were considered to be bombs in waiting. An NPP is not a bomb, but there are risks associated with the release of radioactive materials. In fact, explosions are more likely to occur in chemical plants and conventional power plants. In the past, failures that would release the entire contents of the core into the air and distribute that in the worst possible way were postulated. In practice, multiple barriers exist to prevent the release of large quantities of radioactive materials that would make such a scenario unlikely. One may think of the Chernobyl accident in which a large amount of radioactive material was released. In the case of these types of reactors, there was no reactor containment as there is for pressurized and boiling water reactors. More about the accidents involving Chernobyl and Three Mile Island (TMI) are presented in Chapter 8.

The nuclear industry in the beginning was mainly concerned with the release of large amounts of radioactivity after an accident and therefore concentrated its concern on the possibility of an uncontrolled increase in power leading to core melt. Control and protection systems were designed to rapidly reduce nuclear power generated from the core by the insertion of neutron absorbers. These were in the form of absorbers rods dropped into the core or by the injection of fluids to perform the same function.

The industry has addressed the issue of NPP accidents by designing safe plants with redundant protection systems and a series of barriers to prevent the release of the radioactive elements into the air. The analytical processes used to study the safety of the plants have developed over time, along with an understanding of the distribution of risk contributors. What were once thought to be the main contributors to risk have been replaced by other contributors. This does not mean that the plants are less safe than before; in fact, they are much safer because of the deeper understanding of where the risks really lie. For example, it was once thought that the release of radioactivity because of reactor vessel failure was considered to be a credible event, and now it is not ranked very high on the list of accident precursors identified as risk contributors due to steps taken to avoid vessel embrittlement due to neutron bombardment.

The idea of a very large accident caused by the failure of the reactor vessel and the failure of containment now seems unlikely, but it was once accepted as a design basis scenario. However, it was reexamined from a different viewpoint than that of risk–benefit. This accident was originally thought to be credible. It led to methods and procedures to limit the effects of the vessel failure and to capture the core debris (a core catcher). Another consequence was to design of redundant mechanisms for heat removal of steam release to prevent the structural failure of containment. Because of the possibility of hydrogen being released as a result of a nuclear accident leading to an explosion, a controlled method for burning the hydrogen was introduced into the containment and activated after an accident.

A review of some other accident scenarios showed that they were more likely to occur, and the consequences could be significant. The risk of these "lesser" accidents was in fact higher in retrospect. Core damage could occur, like with TMI Unit 2, but the wholesale release of radioactivity is extremely unlikely. Nuclear plants not having containment does increase the risk of releasing radioactivity to the atmosphere.

The initial safety studies focused on what was called "design-basis" accidents. These accidents were selected to test the power plant designs to ensure that the plants were safe to operate and helped specify the requirements of the plant protection systems. A variety of accident scenarios were considered in this process, including accidents that lead to loss of cooling fluid, an uncontrolled reactor excursion, loss of circulation, and loss of heat removal. The designers looked at ways in which these things could occur and developed the design-basis accident concept. One example of this was the large loss of coolant accident (LOCA) due to fracture of the reactor pressure vessels. There were, of course, other accidents that were less severe, which were considered as part of the suite of accidents to be considered. The outcome of these accidents was the design of the reactor plant and control and protection systems to ensure that the reactor was as safe as possible. In the case of the design of the safety protection systems, the concept of "single failure" was considered so that the probability of a safety system failing to perform its function was reduced. So the industry designed systems, including fluid systems to inject fluid and to cool the reactor, that would continue to perform their functions even in the presence of failures. The concept of periodic testing was introduced to try to ensure that no more than one failure was present in a given system when the plant was operational. Later, the concept of common mode or common cause failures was developed from experience with some early reactors at Oak Ridge National Laboratory in Tennessee.

The safety of the reactor is ensured by having a set of static barriers between the reactor fuel and the outside world. The fuel pellets are enclosed in tubes or rods made of metal cladding. This forms the inner barrier. The fuel rods are then placed within a vessel called the reactor vessel. This is the second barrier along with the fluid circuit that is part of the heat removal system. The third barrier is the containment structure. These reactor systems are pressurized and boiling water reactors. There are other designs that have different arrangements to enhance their safety, but there are fewer of these designs. As seen in Chapter 8, the difference between having a containment or not can be crucial. The cost of a severe accident for a plant with containment is money, whereas an accident for a plant without containment or with limited containment is both loss of money and loss of life.

In the early days all that was thought to be necessary to be safe was achieved by dropping control rods into the core space to shut the reactor down and cut off the heat being generated as a result of the nuclear reaction. However, it is not that simple. A nuclear reactor is not like a gas-fired plant where when you switch off the fuel the heat stops. Decay heat is generated as a result of the nuclear reactions, and this means that heat must continue to be removed after the core is shut down. This means that systems that continue to remove a significant amount of heat for a long time are necessary. The initial heat released is about 10% of the peak power, and drops off over time. Decay heat calculations are contained in industry standards developed by the American Nuclear Society (2005).

The industry started to reexamine the influence of various nuclear accidents over time and steps were taken to perform meaningful analyses relative to understanding the risk and the role humans play in those analyses. Many of the analytical processes used to study safety have come from the nuclear industry and can be used in other industries. The idea of risk, as compounded of both probability of an accident along with an estimate of the consequence of that accident, developed in a major way as a result of looking at operations associated with NPPs. A significant step was taken by the USNRC to sponsor the first probabilistic risk assessment for PWR and BWR plants, and the results of the study were published as the Rasmussen report, or WASH 1400, in 1975.

WASH 1400 started to affect the thinking of power plant design and operation. It pointed out that the concepts behind the design-basis accident were acceptable as far as they went, but there were many more things to consider than just the design-basis accidents, which may or may not occur, such as other postulated accidents that may occur. Thus the safety of the plant is judged more by the response to those postulated accidents than to the design-basis accidents. The design-basis accidents led to a focus on core integrity. The safety systems were focused on how to ensure the core is covered and heat is removed. The possibility of a return to power was obviated by the use of nuclear poisons, such as boron, added to the coolant. Then, to cover the aspect of removal of decay or residual heat, there were systems called residual heat removal (RHR) systems for that purpose. However, it turns out that the core is connected to the secondary side (steam generators for PWRs) for heat removal and uses an auxiliary feed flow system with the main feed flow used for power generation purposes. Features like pump seals are cooled by component and service water systems, and then there is finally an ultimate heat sink (sea, river, or atmosphere). The reactor and plant requirements are more complicated than just requiring a set of shutdown rods.

It also appears that the functions that the main control room crew have to contend with are more complicated than just increasing or decreasing power and occasionally backing up the automatic protection systems should they fail. The PRA studies revealed a complex picture of the integrated systems supporting power generation and ensuring that the reactor core was protected at all times despite the various ways that failures could lead to accidents. In addition, the role of humans in this process was originally underestimated. It was thought that it was possible to design automatic systems to cover any potential accident. Also, the criteria for operation of the plant were increased to include not only accident prevention and mitigation, but also economic operation of the plant. This includes accident avoidance that leads to the loss of equipment and the loss of power output. These criteria are significant for the utilities operating the plants, but they were not part of the licensing requirements for the plant, even though safety and economic operation are intermingled. PRA is an extremely useful tool if used correctly. One needs to reflect on the operation and design of the actual plant as opposed to a generic version of the plant. This has some significance when it comes to considering the impact of humans on safe and economic operation of all kinds of plants, including nuclear.

3.1   LICENSING REQUIREMENTS

Nuclear plant licensing requirements in most countries called for the assessment of the safety of NPPs. The initial approach was the design-basis accident, the defense in depth concept, and the single-failure criterion. This approach was useful for designing the plant, but no assessment of the risk of operation was taken into account. Later a question was raised in the case of potential accidents: What is the probability of an accident other than a design-basis accident occurring, and what is its consequence? A simple definition of risk is a multiple of the probability of an accident occurring times the consequence of that accident. Because the consequence of an accident might be severe, one wanted to ensure that the probability of occurrence was small—that is, the risk was tolerable and comparable to or lower than other industrial risks. The current method for assessing the risks of high-risk technical systems is the probabilistic safety or risk assessment, PSA or PRA.

The PSA is a logic structure that is used to relate accident initiators, equipment, and human success and failures and consequences together in a logic array through their probabilities. The logic array represents the various combinations of possibilities that may occur. Pathways through the logical array represent various combinations of successful actions and failures and lead to certain consequences. A range of consequences may occur from no effect on the plant to a severe impact leading to reactor core damage, release of radioactivity, and deaths of both plant operation personnel and nearby citizens. The PSAs/PRAs are decision tools for management and are not just limited to the nuclear power industry. They can be used in other industries and can be used to focus attention on other operational choices (see Frank, 2008).

3.2   PSA LOGIC STRUCTURE—EVENT TREES AND FAULT TREES

The logic array is made up of two kinds of logical structures called event and fault trees. The event tree represents the series of events that can occur during the accident sequence. It consists of the initiating event (IE), a series of events such as turbine trip, reactor trip, charging system operation, residual heat removal, and, of

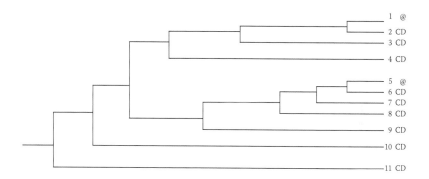

Medium LOCA	Reactor Trip	ACCUMUL-ATOR	HIGH PRESSURE INJECTION	CONTAINM' SPRAY INJECTION	RCS DEPRESS-URIZATION	HIGH PRESS-RECIRC-ULATION	LOW PRESS. INJECTION	LOW PRESS. RECIRC.	CONTAINM. SPRAY RECIRC.	#	END-STATE
IE MLOCA	RT	ACC	HRI	CSI-ML	OLI-ML	HPR	LPI	LPR	CSR-ML		

FIGURE 3.1 Typical event tree (medium loss of coolant accident).

course, the consequence of the set of events that occur. Figure 3.1 shows a typical event tree.

Although the NPP designer tries by a combination of methods to terminate or at least mitigate the consequences of accidents by the use of equipment redundancy and diversity, there are cases when the operator is called upon to act to achieve the same objectives. Thus operator success and failure also has to be taken into account in the event trees. The events in the event tree can cover human actions and equipment failures. In events covering equipment or systems, fault trees model the details of the interactions. Failures of systems can be due to a number of causes, including material failure, loss of power, and human actions. Figure 3.2 shows a typical fault tree.

Humans can cause accidents to occur. Most PSAs consider such things as accidents being caused by equipment failure, such as pipe ruptures due to corrosion, as well as the possibility of human actions that can lead to accidents.

3.3   HUMAN ERROR CATEGORIES

This section covers the four main human error categories along with some discussion of dependence between human activities. Human errors are sometimes thought of as spontaneous errors made by individuals and crews; however, most errors are induced by the situation under which persons operate. They may be caused by technical or human factors in procedures, by misinterpretation of statements within the procedures, by failures in training, and by instructions generated by management.

3.3.1   HUMAN ERROR TYPES IN PRAS

In a PRA there are four main categories denoted by types of human error involved in different plant processes. These have for convenience been called types A, B, C, and

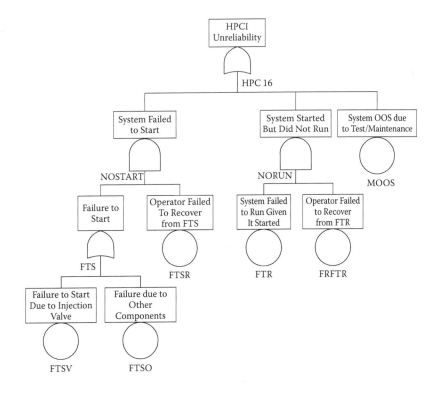

FIGURE 3.2 Typical fault tree.

C_R. These types cover the actions due to various personnel within the plant. Some of the human actions are accounted for within the plant logic tree, some are accounted for in the event trees, and others are accounted for in the fault trees, in the initiators, and in some branches of the event trees. Figure 3.3 depicts the locations of the various human actions in a simplified logic tree.

The root causes for these actions may stem from a number of issues, including equipment design, layout, material selection, control board design, procedure layout, training aspects, and management instructions. Here one will deal with the direct relationship between the action and the personnel carrying out that action.

Actions resulting in type A actions, sometimes called latent failures, or preinitiator actions, can result from maintenance and test operations; therefore the persons associated with these actions are maintenance and test personnel. Because plant operations personnel can also be associated with the organization of the plant configuration, latent errors may also result from their activities.

Type B actions are those that result in an accident initiator. In full-power PSA studies, it is usual to subsume human actions along with equipment failures that give rise to accidents. However, in low-power and shutdown (LPSD) PSA, it is more common to consider human actions that can lead to accidents. Accidents can result from the actions of maintenance, test, and operational personnel. For example, an electrician replacing a particular module may drop a tool, causing a short leading to

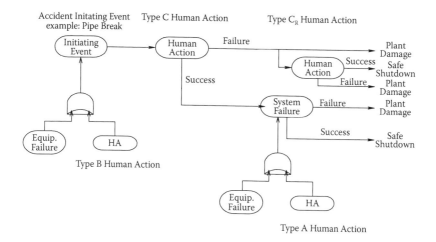

FIGURE 3.3 The location of human error types within the plant logic tree.

reactor trip. In more severe cases, a crane operator moving a refueling cask may drop it, leading to reactivity release.

Type C actions are the expected normal responses of the control room crews to accident initiators. Some analysts call these actions recoveries or postinitiator actions, meaning those actions taken by the NPP crews to terminate the accident or mitigate the effects of the accident. The actions of the crews are controlled by abnormal and emergency operating procedures (AOPs and EOPs). The crews are also influenced by training, time spent in simulated accident situations, and their plant experiences. Their attitudes may be affected by the policies and instructions presented to them by their management. Examinations called for by regulations and industry guidance may also have a strong influence on their responses. In carrying out their responses to accident situations, erroneous actions may result. The cause of these erroneous actions may be derived from the same procedures due to incorrect interpretation of the procedures. Equally, errors may result from the way that information relating to the accident progression is portrayed to the crew, due to human factor deficiencies in the man–machine interface.

In the PSA and in real accident mitigation events, the station crew does not rest purely on the actions taken by the crew within the control room. There are other staff members available to take actions, under the direction of the control room staff and plant management. These actions are called type C_R operations, or recovery actions, outside the precise guidance of the EOPs. The staff used under these conditions includes auxiliary operators, maintenance staff, and other personnel. These actions may include operating pumps locally, closing switch gear locally, and operating equipment that fails to start and run from control room operations. Of course, the station personnel may fail to start the equipment through lack of knowledge or experience, or they may use the wrong steps in the process, leading to human errors during recovery events.

3.3.2 Dependence Between Various Human Actions

One element that can quite easily be neglected in the study of the effect of human actions upon the risk of power plant operation is the possible dependence between human actions in accident sequences. Swain and Guttman (1983) among others have addressed this issue. Human actions appear in both event and fault trees. The distribution between the event trees (ETs) and fault trees (FTs) depends on the philosophy of the PSA group. Dependence between human actions is easier to appreciate in the case of ETs. As can be seen from the ET example in Figure 3.1, human actions may cover detection, diagnosis, and decision making as well as taking action by operating switches and controls (Figure 3.4).

The control room crew's activities are modeled by cognitive actions of detection, diagnosis, and decision making (D^3), and a set of manual activities (M). Often the shift supervisor or the crew chief undertakes the cognitive activities of detection and decision making (often with the help of procedures), and the control board operators undertake the other activities. In this case, the success of the board operators is dependent on the supervisor. Equally, if one operator takes the board operations, his success in performing the second operation somewhat depends on his first action. In the HRA, the process is to examine whether there are cognitive connections between the determination of the accident and actions to be taken. One can see this in the case of Figure 3.4. Dependency exists when the same person takes the actions. It is assumed in most PSAs that the degree of dependency reduces as the time between one action and another action increases. During long-term activities, such as long-term heat removal, the crew changes or other personnel (shift technical advisors) appear to bring other views and perspectives. This can act as a way to break the cognitive lockup of the original crew. A good example of this is the TMI accident. A supervisor from the Unit 1 reactor came into the control room of the Unit 2 reactor and after a while stated that they were in the middle of a LOCA. These are the main effects that should be considered in the various dependency models. More details about dependency models that have been used in various PSAs are addressed in Chapter 6.

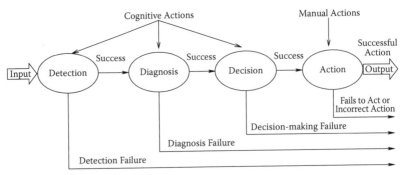

FIGURE 3.4 D^3 M model of control room operations.

3.4   ORGANIZATION OF HRA STUDIES

A key part of the PRA studies is the HRA portion. It appears that people focus mainly on the HRA modeling of human actions, but over time the realization has taken place that HRA personnel ought to be involved in the PRA development process. Swain at one time said how he was utilized during the WASH 1400 by the PRA analysts. He said that he was isolated from the rest of the PRA personnel and it was like being in a room with a hatch. At times, the hatch would open and a piece of paper was passed through with a description of an operator action and a request to quantify the action. He then quantified the action and passed the result back through the hatch. Of course, he was having his fun with us. This is an unacceptable way of approaching the quantification of human events.

3.4.1   Human Reliability Analysis Process

Human error has a substantial effect on the reliability of nuclear power generating stations and other industrial plants. In general, the human contribution to overall system performance is considered to be more important than that of hardware and software reliability. To achieve a precise and accurate measure of nuclear power generating station reliability, human error must be considered. Estimates of human error probability and the resulting consequences of the human error are the principal products derived from performing HRA.

HRA is an iterative process for determining the probability that human actions required for system performance. HRA results are often used as inputs to PRAs that analyze the reliability of entire systems by breaking each system down into its constituent components, including hardware, software, and human operators.

The HRA is a predictive process that determines the contribution of human errors to predetermined significant system failures. The HRA treats the relevant human actions as components in system operation and identifies error probabilities that could significantly affect system status. The main result of the HRA is a set of estimated plant and situation-specific human error probabilities. These error probabilities are incorporated into the total PRA on the basis of their effects on the reliability of a component, an entire system, or the entire response scenario required by an initiating event.

The recommended practice for conducting HRA is based upon the IEEE Standard 1082, "IEEE Guide for Incorporating Human Action Reliability Analysis for Nuclear Power Generating Systems" (1997). As indicated in IEEE Standard 1082, the HRA process "parallels the typical PRA process but is not organizationally related to it." The process should be "compatible with the PRA" and is the method for "incorporating human actions into the PRA." The timing and level of detail of the HRA process "must match the PRA."

It should be noted that the HRA process given here has been widely adopted within the nuclear power industry and has been applied by other industries. However, the process does not select a specific model or models for the quantification of the human events considered within the PRA plant logic model. So the process can be applied by different persons with different needs and requirements.

The ASME PRA (ASME, 2002) document is the current standard for PSAs/PRAs in the United States, and it has identified three general analysis categories that lend themselves to the needs and requirements of the industry. It is suggested that the standard be reviewed by any organization's PRA group for guidance. A similar process is suggested for the HRA. It is asserted that in the HRA domain there is no absolute best practice. The best practice for one circumstance is not necessarily the best practice for another circumstance. This concept mirrors the viewpoint of the ASME standard. The selection of a model of human success or failure depends on the precise needs of the organization.

Organizations may have different needs for a PRA. In one case, the organization may wish to take advantage of improvements in its training program to consider crew manning changes. In another case, the utility needs only a gross estimate of plant probability of damage to various systems. Thus the selection of a HRA model along with the associated data depends to large measure on the requirements of the organization and the overview of the corresponding licensing authority.

Numerous HRA methods available to evaluate the contribution of human error to plant risk are discussed in Chapter 5. The ASME standard indicates three capability categories and a breakdown of requirements for pre- and postinitiator HRA interactions. The ASME standard states that for each category, a utility may claim that the analysis more effectively reflects the specific plant from a safety viewpoint. This means that the utility can claim relief from USNRC requirements, such as reduced in-service inspection. Other organizations could use this kind of approach to satisfy their safety needs and meet those of the corresponding licensing authority.

The HRA process model presented in this chapter is a structured, systematic, and auditable process that provides assurance that the reliability of risk-important human actions is accurately estimated so that the effects of these actions can be assessed, using PRA.

As shown in Figure 3.5, the HRA process model consists of eight steps. Steps 1, 2, and 3 are the "initial planning" steps for the HRA. These steps consist of selecting the appropriate team members to conduct the HRA, training them to perform as an integrated team, providing the team with the required plant-specific information needed to conduct the HRA, and constructing an initial plant model that integrates human performance considerations with the hardware and software considerations. Steps 4 through 8 are the HRA implementation steps. These steps are screening, characterizing, quantifying human interactions; updating the plant model; and reviewing the results of the HRA in the context of the PRA.

The eight steps of the HRA process are presented in the following sections in more detail.

3.4.1.1   HRA Process Step 1: Select and Train HRA Team

The first key step in the HRA process is to select the HRA team. The HRA team leader needs to be experienced in the HRA process, be able to identify HRA models and methods suitable for the needs of the utility, be knowledgeable of PRA, and be knowledgeable about the requirements of the regulatory authority. The training of support staff can be undertaken by the team leader as part of his or her duties. The HRA group will interface with the PRA analysts and PRA team leader. The

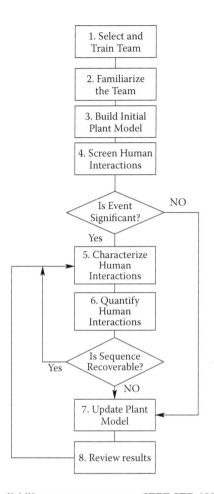

FIGURE 3.5 Human reliability assessment process (IEEE STD 1082, 1997).

HRA leader should be well trained in the use of expert elucidation techniques—that is, well versed in the use of expert judgment for human error probability (HEP) estimation purposes. The quality of the HRA depends on the HRA leader. It is simply not enough to direct a person to use a given HRA model without understanding its limitations.

In carrying out an HRA study that reflects the plant and personnel, there is a need for plant domain experts to be either attached to the HRA group or be available to give expert advice to the HRA group. The domain experts could be instructors, maintenance and test personnel, control-room crews, and other operational personnel associated with incident analyses.

The HRA persons need to be trained in the characteristics and use of procedures, control room layout and functions, and crew structure and responsibilities. They need to have an understanding of maintenance, test and calibration methods, and procedures, including work orders and task allocation between personnel.

3.4.1.2   HRA Process Step 2: Familiarize the HRA Team

Given the background knowledge of plant operations achieved by training, the HRA personnel should become familiarized with the plant and its operation. This can be accomplished by observing simulator training exercises, classroom activities, and various maintenance activities. HRA persons do not have to be domain experts but should understand the jobs undertaken by these various persons.

3.4.1.3   HRA Step 3: Build Initial Plant Model

The building of the initial plant model is seen as a function primarily of the PRA systems analysts. Often the analysts repeat logic structures of successful earlier PRAs. However, the HRA team should consider developing event sequence diagrams (ESDs) to review the set of human actions that can lead to plant or core damage along with equipment failures. Insights from the ESDs can lead to improvements in the plant logic model. The ESDs are derived from an understanding of the transient behavior of the plant and the interactions with both correct and erroneous human actions.

3.4.1.4   HRA Step 4: Screen Human Interactions

The function of performing this step is for the team to ensure that the HRA resources are spent in the most cost-effective manner possible and that each human action within the logic array is accounted for in terms of its contribution to plant damage or core damage frequency (CDF). To ensure that this role is met, each screening value is given a conservative value in the plant model. The initial plant model is then inspected to determine which human actions are important, and these actions are then listed for more precise evaluation. The remaining human actions are left with their conservative values. Once the more detail evaluation has been carried out, these values are embedded into the PRA, and the case is rerun. This is discussed in more detail in Step 7.

In general, criteria need to be established for a PRA application to conduct screening. The objective of HRA is to identify and assess the impacts of human performance on system reliability as realistically as necessary. This does not necessitate that all analyses be as realistic as possible, as this may lead to unnecessary analyses and expense. For some PRA applications, a more conservative (i.e., less realistic) treatment of human performance may be more than sufficient to meet the needs of estimating an upper bound for certain classes of human failures or categorically limiting the types of human failures in further analyses. The burden in such analyses rests on the ability to demonstrate that the range of likelihood (for quantitative screening estimates) and kinds of human failures (for qualitative screening) are bounded by application of the screening criteria. For example, a review of certain component unavailability rates may indicate a per-demand failure probability in the range 5E-2 to 1E-3, including the contribution from all causes (i.e., fatigue related, aging, human failure induced, etc.). Using this information, a screening range can be applied for human failures with these components, assuming appropriate uncertainty regarding the distribution of the failure rate. Because human contribution to system unavailability cannot exceed the rate of system unavailability due to all causes, an analyst may use such information to limit the range of human failure probabilities within this system. For purposes of screening, such criteria are effectively conservative and limit the resources at this stage of analysis that need to be applied to certain human failures.

For screening human actions in the PRA, a set of conservative screening values for HEPs should be used. The values depend on the Human Error or Human Interaction (HI) types—that is, types A, B, and C, and the modeling of the action within the PRA logic model. By the latter is meant whether the HI is holistic or atomistic—that is, a given HI appears as one or more values in the logic array.

Depending on the modeling structure that the HRA analyst selects for the human action and its location within the ETs and FTS, the following conservative values may be used for type A and type C human interactions 0.01 and 1.0. In cases where the type C action is divided between the cognitive and manual actions, the cognitive value could be set conservatively to 1.0 and the manual value set to 0.1. If the HRA tree approach of Swain is used, but conservative values are used, repeated values set to 0.1 can lead to unacceptable low values for manual action errors. For example, four such operations can lead to an action HEP of $0.1 \times 0.1 \times 0.1 \times 0.1 = 1.0E\text{-}4$, which is hardly very conservative.

3.4.1.5   HRA Step 5: Characterize Human Interactions

This step is to examine each of the important human actions from the point of the context that affects them. Human errors can be characterized into two main groups: random errors and situational-dependent errors. From the point of view of PRA, one is only interested in those errors that are adjudged to be systematic errors. The first group, random errors, is unpredictable and uncertain. The second group is dependent upon the situation or context as it affects all of the personnel. In other words, errors derived from individual errors should not be used in a PRA, because the probability of an individual making an error at the same time as an accident is low; that is, the joint probability is so low as to preclude consideration. In the case of an accident one should ask, is the operator action symptomatic of the crews as a whole or is this just an individual issue?

In part, the characterization of each human action depends on the requirements of the PRA, so if the need is for an approximate representation, then the details will be limited. If, however, the PRA requirement is for a very detailed HRA representation, then the model selected will reflect this. Chapter 5 gives a number of HRA methods, and each method sets up the requirements for the characterization of the human action. So if THERP is selected, then a task analysis of the action has to be carried out in terms of what are the indications, alarms, diagnoses, decisions, and manual actions. Also, the environments under which the actions are taken have to be understood so as to evaluate the impact on the human performance, such as stress. For each human action in an accident, there is a context under which the personnel involved are operating. The interplay between the human action definition, the operating environment, and the HRA team's selection of the HRA model determines the needed characterization of the human interaction. Of course, the operating environment is affected by the accident being considered; for example, specific alarms and indications are activated during the accident and the effect of these may be different from another set of alarms or indications for a different accident, and they need to be accounted for—that is, the context may be different for different accidents, and therefore the human error probability will also be different.

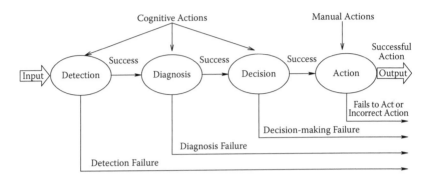

FIGURE 3.6 Behavioral model.

The characterization also covers how the operational personnel are considered within the PRA/HRA structure. For example, control room crews are composed of a number of individuals, each performing separate tasks when responding to an accident. One could consider modeling a crew as separate individuals fulfilling separate tasks but coordinated or as an entity. It is difficult enough trying to provide data for even simple models of a crew, so it is advisable to consider the crew as an entity.

Behavioral models have been suggested to model individuals but can be considered to model the attributes of a crew. Such a behavioral model is depicted in Figure 3.6.

The various behavioral aspects of the model are fulfilled by the crew members. Detection is really a function of all crew members, but the control board operators (or equivalent) perform a monitoring function. Diagnosis and decision making are functions of the procedure reader, with an overview function performed by the shift supervisor, and additional cognitive functions performed by the shift technical advisor. Once a decision has been made, the board operators are expected to carry out the actions by operating controls and switches.

This model is introduced here because some analysts characterize the cognitive actions and manual actions differently within the plant logic; for example, the cognitive aspects of detection, diagnosis, and detection are modeled within the ETs, and the manual actions in the FTs. This type of split-up has implications in the later HRA processes of screening, HRA model selection, and quantification and consideration of dependency between human actions. As mentioned earlier, HRA is an iterative process; there are connections between selections of the HRA models, characterization of the human actions within the plant logic, and availability of data (plant, simulator, and expert judgments). For example, the availability of data may dictate the selection of the HRA model or the behavioral characterization of the crew as depicted in Figure 3.6.

The output from this task is a list of the human actions to be modeled and quantified.

3.4.1.6 HRA Step 6: Quantify Human Interactions

Human interactions have been screened, the plant logic model has been examined, and the dominant sequences have been identified along with the corresponding human interactions. These actions now have to be quantified. They have been identified with respect to whether they are types A or C. It should be pointed out that type B actions are usually subsumed for full-power PRAs within the accident initiator frequencies, although type B actions become more significant for low-power and shutdown PRAs. Type C_R actions are evaluated after the main PRA is complete and the results are known and the organization is interested in seeing what further actions might be taken to reduce the CDF value—that is, how to make the plant safer.

The key human actions have been identified, at least for conservative HEPs. The HRA personnel are now faced with reviewing the utility needs for the PRA, availability of plant domain experts, simulator data availability from accident scenarios, context definition for each accident scenario being considered in the PRA, data from past plant incidents and accidents, and insights from near-miss incidents, which could include write-ups or access to plant staff who witnessed the incidents. Based on the above data and the experience of the HRA leader and others, selection of an HRA model or models would be made. A number of available HRA models that could be selected are listed in Chapter 5.

3.4.1.7   HRA Step 7: Update Plant Model

This is an iterative step in that the plant model can be updated a number of times, starting with changes following the use of screening HEPs values. As mentioned above, the screening HEPs are inserted into the logic trees in their appropriate places, and the sequence core damage frequencies are calculated. Equivalent measures would be examined for nonnuclear applications. Following the review, the key human actions are quantified based on the selected HRA models, are inserted into logic trees, and the sequences are recalculated. The results are reviewed to see if the order of the sequence's CDF contributions is changed so that a sequence for which HEP is based on conservative screening values rises to a level of being a dominant contributor. In this case, the action is requantified based on the selected HRA model. The process is repeated until the order of key HRA contributors remains unchanged.

3.4.1.8   HRA Step 8: Review Results

Reviewing the results of the PRA should be carried out by a number of persons for different purposes. The first review is to see if it is necessary to extend the scope of the HRA as carried out to include recovery actions. As mentioned before, these actions are those that can enhance the safety of the plant but might be considered extraordinary. For example, the manual operation of the steam turbine-driven pump may be considered to be extraordinary by some but within the procedure scope by others. So, in one study it is included and another study excluded. However, faced with a need to reduce the risk of plant or equipment damage, it might be included, and the action of the utility would then be to examine the EOPs so as to include this feature. The HRA would be modified to include evaluation of this action based on acceptable HRA methods covering field operations.

This review should be considered as part of a required internal review by utility personnel. For these reviews, the utility would call upon the service of knowledgeable operations personnel as well as the PRA project team.

In addition to the internal reviews, it is recommended that the PRA/HRA be reviewed by outside experts to see that it conforms to good practice. The outside experts should examine the bases of PRA/HRA, the selection of models and data, and the interpretation of plant and simulator data. The application of the models and HRA calculations should be reviewed and discussed with the appropriate team members. The outside reviewers would list any areas found to be questionable and request answers from the PRA/HRA team. Based upon their work and the replies to questions, the review team would prepare a report for the utility managers.

A report on the PRA would be sent ultimately to the licensing authority, the USNRC, for review. However, the onus for safe and economic operation of the plant lies with the utility management.

It should be pointed out that although the above HRA process was developed for NPP PRAs, it can be applied to other high-risk industries or operations. The process has been applied by National Aeronautics and Space Administration (NASA) in some of its projects, such as the shuttle PRA and the International Space Station PRA.

4 HRA Principles and Concepts

INTRODUCTION

Human reliability is a central issue to the estimation of the factors that affect safety and availability. At the time of the first PRA studies (i.e., WASH 1400, 1975) and AIPA (Fleming et al., 1975), the importance of the HRA contribution was underestimated and it was thought to be about 15% of the total risk. HRA is currently estimated to be 60% to 80% of the total risk. The change is thought to be associated with increased understanding of the human contribution, but steps have also been taken to reduce the equipment contribution to risk by better maintenance, better selection of materials, and better designs. Over the years, organizations and individuals have spent a great deal of energy trying to better define the key HRA issues and have been proposing methods to predict and quantify the issues associated with human performance deficiencies in various situations. This chapter introduces a number of elements that affect HRA formulation.

4.1 BACKGROUND

There has always been a problem with understanding human reliability and what kind of academic topics form a useful background for understanding HRA. What is HRA? What are the theoretical elements of the methods advocated for its use? HRA is the process of assessing the probabilities of human error under certain circumstances, namely, accident conditions. There are a number of scientific areas associated with HRA, including human factors, psychology, and accident analyses, but HRA has developed into a topic that stands on its own.

Some people tend to group HRA within human factors, but human factors are really, in the opinion of the author, a design tool. Psychology covers certain phases of human performance that can be seen in HRA, but HRA is more of an engineering tool to be used to study the risk of operating, and in the design of high-risk industries, as a social tool to analyze people. Psychology deals with the human state and is divided into a number of subdisciplines. One of these is cognitive psychology, and this topic covers some points of interest to HRA experts. Cognitive psychologists are interested in how people understand, diagnose, and solve problems, concerning themselves with the mental processes that mediate between stimulus and response. HRA specialists try to capture these elements within the HRA formulation.

The reliability of an individual is subject to a large degree of uncertainty and is more appropriately a topic for psychologists. Industry is in the business of trying to reduce uncertainty and variability in their operations. There is a continuous debate about what the proper roles are for man and machine. The line between man-operated systems and automation has continuously varied. In the early days of nuclear power, safety equipment was thought to be acceptable for dealing with a range of accidents called design-basis accidents. Later, it was seen that this was not so, and the role of humans became more important to safety; hence there was a drive toward a better understanding of human reliability and the things that can affect it. The reliability of humans can be enhanced by forming a team, so that one member supports another and can augment the others' knowledge and experience. Additionally, one can cover some of the weaknesses in terms of memory and knowledge by the use of procedures. Procedures are the repository of the knowledge and experience of persons who can think through the best response to an accident. The procedure designers can accomplish their work under nonstressful conditions and can then test their procedures against accident codes and full-scope simulators. Industry then requires the crews to be trained both by book and by simulator.

Accident analysis is also closely aligned with HRA in that the study of accidents can lead to a better appreciation of what aspects in the organization of plant operations can influence persons operating a plant or equipment. The study of accidents can help the HRA analyst clarify the potential influences between plant components, personnel, and management controls and direction. It is all too easy to focus on what appears to be an issue, only to find out that this is too narrow a view of the situation. In performing a PRA, one may decide that maintenance operations are limited to just the availability of a pump or valve, whereas there may be critical interactions between operations and maintenance.

Studies of crews during power plant simulations are effectively studies into the execution of personnel responses to accidents under conditions that can be observed and recorded. Here rules can be written for the failure of components, and initiating events can be selected so that the simulated scenario can credibly represent accident scenarios that could occur in practice. In the case of real accidents, initiating events are random and often infrequent, thank goodness. The objective in the simulated case is to see how the operators respond and how operator aids can help ensure that the plant is returned to a safe state. In carrying out these studies, the limitations of the training of personnel, procedures, and human–system interfaces, can be evaluated.

Often in the case of accidents, records are in short supply. The aircraft industry has possibly one of the best approaches to accidents in that they have flight recorders, which can outlive the consequences of accidents. These recorders have records of a number of airplane variables, such as speed and altitude, and they also have records of the key variables that affect the aerodynamic characteristics of the plane, such as elevator, rudder, and aileron angles in time. Engine characteristics are collected such as thrust measurement and fuel supply variations for each engine. A number of measurements are taken that relate to specific instruments, so that one can see what information was presented to the pilots along with control information related to the actions taken by the pilots (i.e., control stick or equivalent data). Voice recordings are also collected and stored in the flight recorders. Accident investigators have to enter the minds of the pilots aided by this information, since in many cases the pilots fail to survive the accident along with passengers and other crew members.

Alan Swain and his associates tackled the question of how to assemble a nuclear bomb reliably so that it would function only when required and not at any other time. This was a significant task, and they developed a valuable method for the estimation of the reliability of the operation. Others, such as the Russians, British, and French, were faced with the same issue. They most likely tackled the problem in much the same way. Swain was asked by the U.S. Nuclear Regulatory Authority to assist the Rasmussen PRA group to use his HRA techniques on the landmark PRA study, WASH 1400, in the period before 1975. Later, the HRA technique developed for WASH 1400 was written up as the HRA handbook and was issued first in a draft form in 1979 and later in modified form in 1984. This was an exceptional piece of work. It covered many aspects that are central to a complete HRA study.

4.2   HUMAN RELIABILITY CONSIDERATIONS

The earlier chapters in this book refer to plant safety and some aspects of the history of development of plant safety and the accompanying topic of human reliability. In this chapter, the concerns are the requirements for an HRA model, the relationship between psychology, human factors, and accident analyses, and the data to use when performing an HRA. The skills and knowledge of the HRA specialist also play a role in the process in that the selection of an appropriate HRA model and the associated data are key measures for the correct interpretation of the risk of operation.

The level of trust that one has in the HRA methodology in general enters the equation, because if one has no trust in the HRA methodology or in PRAs, then its messages may be disregarded. Some people feel that HRA is of little value and cannot predict the impact of various changes to the working environment of humans on reduction or increase in risk. HRA technology is changing, and most would say improving.

Long before the arrival of HRA techniques, people were looking for improvements in man–machine interfaces. Manufacturers supplied information about the states of their pumps, motors, and cars when operating. The information was supplied in the form of gauges and indicators. These gauges and indicators were assembled together on a panel. The gauge sizes were variable, and markings could be difficult to read. There were no standards or rules governing their readability. Because the devices were being used by people, it became clear that they ought to be designed with people in mind. The subject of human factors started in this way. Human factors, such as with the Mil Standards, Mil Stand 1492F, (2003) and NUREG-0700 (O'Hara et al., 2002) are the latest in a series that has filled that need. The formalism of their approach characterizes a number of domains covering things like readability, including layout of procedures. An associated field is addressing the workstation design to fit humans, with chair designs, lighting intensity, and other equipment—the field of ergonomics. Another associated field is usability—using people to review computer-based applications and make judgments about the quality of the interface in its ability to communicate the message of the application designer. Although the techniques of human factors have done well in improving the man–machine interfaces (MMIs), there is still a need to ensure that the MMIs help fulfill the requirement to improve the reliability of the crews in restoring plants to safe conditions. The real requirement of the MMI is to support the cognitive needs of the crew trying to control accidents. There may be a difference between human factor specifications for static and dynamic situations; therefore it might be necessary to confirm that the MMIs meet accident-based requirements by using simulators to test control room designs.

What does the above discussion have to do with human reliability? Most HRA methods place an emphasis on the quality of the man–machine or human–system interface and also on the quality of the procedures along with other effects. The quality of these influences factor into HRA assessments. How they enter into the HRA process depends on the HRA process. In some cases, the influence of MMI and procedures enters via a compensating influence called a performance shaping factor (PSF). There are a number of features covered by PSFs, such as training, communications, and command structure. In other cases, the influence of the MMI and procedures, or other influences directly affect the calculated human error probability (HEP). Clearly, the design of the MMI, procedures, and other influences are seen to be important in human reliability. Clearly there is a close relationship between HRA and human factors, but the one is not the other.

Human reliability assessment is a process that examines human actions in responding to accidents to predict their contributions to the risk of operation of a power plant, factory, chemical plant, or some other high-risk technical system. Usually the vehicle for carrying out such a study is the probabilistic risk or safety assessment (PRA/PSA) process. HRA is a process that takes input about the plant, its staff, and its operating processes and then predicts the reliability of humans operating the plant during specific accidents. Because one wants to do this in general, one has to deal with an average representative group of the operational staff. So the crew represented in the PRA/HRA is an average of the actual crews, neither good nor bad. Accident initiators are random, so the possibility of a specific crew being present at any given time is low. Predicting which crew might be present during an accident is very low given crew rotations and crew members on leave or out sick. Also, there is variability in the ways that crews respond to a given accident scenario. The results of a HRA study include uncertainty. This uncertainty covers variations in crews' performance and the impact of variability in influence effects or PSFs on the basic HEP value.

Psychology sheds some light on the responses of crews during their accidents and even during normal evolutions. Norman (1988) and Reason (1990) cover the separate actions of slips and mistakes. They differentiated between these two acts by operators. A slip is when an operator goes to operate a particular switch but selects a different but similar-looking switch. In other words, the operator had the correct intent but executed the wrong action. In the case of the mistake, the operator had the wrong intent, and it is almost immaterial which action was taken. These are the type of things that have been observed in practice and during simulated accident scenarios.

Operators seem to recover more quickly from slips than they do from mistakes. The HRA approach should recognize the presence of these different error types.

Mistakes are more important than slips, because they are caused by the operator having the wrong mental model of the accident. In the case of a slip, the feedback to the operator and the crew can be much quicker. For example, if a switch is to be used to operate a pump, then the operator should be aware of an increase in current for electric pumps and an increase in flow from the pump, so recovery is much easier since the operator is prompted. In the case of a mistake, the operator is left tackling the wrong problem and things are not so clear. Certainly the plant response is not necessarily related to the operator's mental model. In cases like this, the operator starts to look for other reasons why things are not going well, such as a failure of instrumentation. This situation leads to "cognitive lockup," so the operator really needs an outside source or opinion to "release" him or her from this situation. This function could be usefully fulfilled by an automated procedure following system (IEEE, 2009; Spurgin et al., 1993). These features should be accounted for in the HRA study. One would expect that the error probability will be much lower during the use of procedure support systems than for the case of using paper procedures. Of course, the design of the computer-aided device has a strong role to play, and some systems will be better than others (see Chapter 6).

Swain produced the HRA handbook (Swain and Guttman, 1983) based upon a task analysis approach with the task elements given HEP values; more about this is contained in Chapter 5. In a review of the work (IEEE, 1979), a number of people criticized the draft document, saying that it did not cover the cognitive characteristics of operators. This led to some changes to the handbook, but the real point here is how we are expected to treat cognitive processing as opposed to manual actions. Following the meeting, the concept of the time reliability curve (TRC) was used in the belief that it represented the cognitive actions of control room crews. Swain introduced a TRC into the handbook to cover the perceived deficiency. Chapter 5 covers this topic in more detail.

What does cognitive processing mean in the context of operator actions and HRA modeling? A simplified model of the various roles played by an individual is shown in the D^3 A model (see Figure 3.4). In this model, there are four modules. The first three modules are detection, diagnosis, and decision making. The fourth module covers actions taken by personnel. This is a symbolic representation of the mind, but the actualities of a crew are more distributed in that some crew members monitor and use the computer consoles or control board elements (displays and controls). These operators act as the eyes and hands of the staff member, who is reading and using the procedures to give instructions to the other staff. In addition, there are other members of the crew, depending on the utilities' crew manning design. There is a supervisor and possibly an engineer, operating as a shift technical advisor (STA). The supervisor performs the function of ensuring that the control of the accident scenario proceeds well; if it does not, then he steps in to supply help to the procedure reader. The STA is expected to be deeply knowledgeable about plant transients. He may be called if things go awry. In some plants, the key parameter that leads to the STA being called is the activation of safety injection. With safety injection being activated, the plant is reconfigured—that is, the reactor and turbine are tripped and the feed water cut back to just the auxiliary flow. In other words, the plant goes into an emergency state. Of course, this may be spurious action, but it is more likely to be a real emergency or unsafe condition.

4.3   CREW MODELS

The central issue with modeling the crew is the view of the PRA/HRA analysts relative to accident progression. If the analysts are more interested in the crew structure and how it works, then the model is based upon models of the interactions between the crew, the control room displays, the procedures, and how the plant transient affects the decisions and actions of the crew. This can present a very challenging task for the modeler; some work on this topic has been carried out by G. Petkov (Petkov et al., 2001, 2004). It also leads one to consider actual crew member characteristics, so it is more of a psychologist's model. This model might consist of multiple D^3A types with connections between the elements, the dynamics of the plant through the MMI, and access to the procedures (EOPs and AOPs). However, for the purpose of a PRA, this seems to be a little too complex, and there is the issue of where the data come from to feed such a model. This does not mean that one should not consider such developments.

One could try to simplify such a crew model so that that D^3A type covers the crew average—that is, it is representative of crews in general. Here we can get a little closer to garnering data by the use of simulators. Even here we find difficulties; because we modeled the three Ds separately, we need information on when the crew detected the onset of the accident condition. The same goes for diagnosis and decision making: When did they occur? Observations of the crews during simulator sessions can lead one to see with specific crews when these separate states are achieved, but different crews have different styles, and this can make things more difficult. These states can occur for different crews at different times. Again, it is difficult to construct a useful model and to get information upon which to make decisions on the overall probabilities of failure of the crews. This is a lot of detail, but it is not a focused view of the data that one needs. It is of interest to note that a USNRC-funded project called SPAR-H (Gertman et al., 2004) made use of a decision and action (D and A) model with fixed human error probabilities for each state (see Chapter 5 for details.

One could decide to look at the success and failure of crews performing tasks to correctly respond to an accident scenario. The accident starts here, and these are the steps taken to terminate the accident. One is not so much interested in the internalization of the operation in the minds of the crew, but rather the success or failure rate of the crews. The problem with this approach is how do you extrapolate the data to other accidents, other crews, and other plants? The virtue of having a phenomenological model is the ability to perform this extrapolation. If we could find a method of extrapolation without complexity and it was fairly accurate, we would not need a detailed crew model. Most of the models proposed so far covered HRA with point models, rather than distributed models. Part of the problem is one of data; it is difficult enough to predict the point model, let alone detailed crew models.

4.4   ACCIDENT PROGRESSION AND HRA PATHWAYS

Seen by analysts considering responses to accidents, the pathway to be followed by the crews is fairly clear. This is particularly true of analysts performing transient safety studies. The analysts know the accident scenario very well, having basically designed it. However, operators of power plants do not know the accident scenario ahead of time. They have to progress by examination of the displays and follow the procedures to terminate the accident. It is possible but unlikely that the operators could make an error at each procedure step. HRA analysts should be prepared to track through the various statements within the procedures to understand the options that the operators might take. Of course, some pathway options may be less likely than others. The analysts can gain information on the likely pathways the operators might take by consulting with training instructors or by witnessing simulator sessions in which the crews are coping with accidents without the assistance of the instructors. In analytical terms, each potential pathway has a probability and an associated set of influences that make the pathway more or less likely.

Several HRA techniques cover the topic of the pathways that might be taken and the forcing functions associated with why these pathways might be taken. Two HRA approaches pay attention to this—A Technique for Human Error Analysis (ATHEANA) (Forster et al., 2007) and MERMOS (Pesme et al., 2007). Others have used hazard and operability studies (HAZOP) (Rausand, 2004) rules to sort through the potential HRA pathways. For example, this approach was used during the Yucca Mountain project—see the License Application report (Yucca Mountain, 2008). Another technique used frequently is the event sequence diagram (ESD). This approach covers sequences generated by equipment failures and human failures. In Chapter 8, a number of these, generated from accident reports, are covered. The ATHEANA and MERMOS focus on human-related activities, but equipment failures can induce human failures (e.g., loss of key instruments can make it difficult for crews to respond correctly to accidents).

4.5   DATABASE REQUIREMENTS AND EXPERT JUDGMENT

As stated above, data are the lifeblood of HRA methods, as they are for most engineering activities. Obtaining data related to human activities is quite difficult. For mechanical equipment, one can perform tests that duplicate the operation situation and note when the equipment fails. In this way, data can be obtained relating the failure probability and a measure of uncertainty. Performing the same kinds of tests with crews responding to a simulated accident only points to the fact that the crews learn by the situation, so their responses change. The other problem is one of carrying out enough tests, especially when the anticipated failure rate is very low. The environments of nuclear plants vary from plant to plant, which makes the idea of performing tests, at all U.S. plants for example, very difficult.

Using simulators can give one an idea of where errors might be hidden because of procedures, displays, and so forth, but simulators do not yield useful human error probabilities (HEPs), except in a very few cases. Typically, there are six crews per unit, and the range of units per station in the United States is from one to three. In other countries, the number may rise to six or even eight. One plant that has carried out many simulator sessions is the Paks NPP in Hungary. Paks has four units or 24 crews, so the lowest probability is 1/24 for one failure. The crews tend to have very low failure rates, lower than 1.0 E-2, when a well-trained crew uses tested emergency procedures.

As will be discussed in Chapter 5, there are different types of HRA models. Associated with each type, there are different methods of determining the HEP for a given situation. Each method needs data in order to calculate an HEP. Some have built-in databases with correction factors, others relate to time with modifiers, and others use anchor values, also with correction factors. All of the models have some reference HEPs or equivalents together with a way of modifying the HEP according to the situation. In other words, the HRA methods have frameworks that help relate the human performance to an HEP. Then in order to ensure that the HEP reflects the actual operating situation, correction factors are introduced.

The model approaches are discussed in depth in Chapter 5. Even with a fairly well-defined database, some correction is required in order to generate an actual situation from a generic situation. How well that is done is a function of the total methodology. If the correction factors are weak functions, then the supposed actual situation is not approximated by the "corrected" generic model. If the influence factors are stronger, then the corrected model may better fit the actual situation. This is where selection of an HRA model is critical. If one is satisfied with a generic HEP value, then the generic model is acceptable. However, if one needs to more closely represent the actual plant, then one should understand the factors that affect human reliability and select a better model. Methods that rely upon experience gained from watching simulator accident sessions can have much to recommend them, if one wants to move to a specific plant-based HRA.

4.6   PRA/HRA ORGANIZATION

The PRA and the involvement of HRA in the PRA need the organization of resources to ensure that the processes are carried out successfully. Both PRA and HRA involve persons with different skills from deep knowledge about plant operations through statistics and human reliability. The need is to gather information on systems, components, plant operations, plant procedures, and training of operational staff and, of course, measurers of their reliability. This information then needs to be formulated into a plant logic structure, which in turn should be embedded into a computer-based logic solver. Over the years, the PRA industry has developed computer tools, such as Riskman (ABS Consulting), Risk Spectrum (Relcon Scandpower, AB), CAFTA (EPRI), and SAPHIRE, USNRC-INEL.

ASME has a document outlining the requirements for PRA, see Probabilistic Risk Assessment for Nuclear Power Plant Applications (ASME, 2002). The EPRI issued two documents dealing with the organization of the HRA portion of the PRA—namely, SHARP (Hannaman and Spurgin, 1984a) and SHARP II (Wakefield et al., 1992). SHARP II is an updated version of SHARP.

More about the organization of the HRA process is covered in Chapter 3.

4.7   REQUIREMENTS FOR HRA EXPERTS

As can be seen from the above discussion, at the center of much HRA model usage is the judgment of the HRA expert. This goes from selection of the HRA model or method, to understanding the relevance of the operating conditions, to the selection of influence factors and their relative weights, and to the selection of basic HEP values. Knowledge about the various HRA models and methods, how to get the best out of them and how best to relate domain expertise to the solution of the HRA problem is key. The difference between a good application and poor implementation lies in the knowledge of the HRA expert and availability of domain experts and how to use them.

5 Review of Current HRA Models and Methods

INTRODUCTION

In this chapter, a number of current HRA models and methods will be discussed from the point of view of their design characteristics. HRA methods differ in their characteristics. Here the characteristics have been separated into three groups: task, time, and context related. The divisions may be a bit arbitrary, but the characterization enables the various HRA methods to be compared against some measures. In Chapter 7, there is a discussion of the pros and cons of each method. This chapter covers a little of the historical development of HRA methods.

All of the developers of HRA methods should be thanked for their attempts to define a difficult topic. The author has laughingly used the Indostan blind wise men trying to define the elephant to point out the analogy to HRA developers—it is not because they are blind but rather HRA is difficult. Like the blind men and the elephant, each HRA developer has captured an aspect of the HRA field. The issue is how to integrate all of these pieces into a meaningful whole. Perhaps it is like the theory of light—the wave theory holds under some conditions and corpuscular theory for other conditions. However, one should continue to look for the unified field theory for HRA.

Often both in this document and elsewhere, mention is made of first- and second-generation HRA models and methods. Dougherty (1990) was the first to write about this as a way of differentiating between certain classes of HRA model. He and Hollnagel among others pointed out the limitations of the first-generation models and the need to have better bases for other later developments, or second-generation models. In some ways, the analogy between first and second generation is like the difference between curve-fitting and phenomenological models.

5.1 BACKGROUND

The topic of HRA has been around in one form or another for some time. Since WASH 1400 (WASH 1400, 1975), HRA has seemed to be tied to probabilistic risk/safety assessment (PRA/PSA). This is not surprising in that reliability has always been tied to a benefit or risk. Even in the case of car reliability, reliability is tied to cost or the failure to provide service when required. So loss is usually tied to a loss of reliability. One early case of human reliability was that of observing, reliably, the launch of Soviet rockets toward the West. It was not just a case of indicating that a rocket was seen to be launched on a cathode ray tube (CRT), but it included the identification, evaluation, and decision process: "Is it really a Soviet rocket launch, is it a misfire, or is it a Chernobyl or some other fire?" The human element was central to the process, and it was a risk/benefit situation with the stakes being very high, up to and including the destructing of a large part of the world.

Prior to WASH 1400, Swain and his associates were engaged in the HRA of nuclear weapons assembly. Here the question was raised would the weapon work when required, not work, or work prematurely. A key element in the assembly process was the role of the human and how it affected the safety of the bombs. The atomic bombs had to be assembled in the correct sequence and order to minimize the chnce of either a premature explosion or a failure to explode when required. The evaluation process used by Swain was the task analysis of the steps taken to assemble the weapon. Swain and his associates then used HRA data or estimates for each step. All of this was formulated into THERP (Swain and Guttman, 1983). This method was released in an early draft document, "Handbook of Human Reliability Analysis with Emphasis on Nuclear Power Plant Applications" in 1979. Following the critical review of the draft document in 1979 at an IEEE conference (IEEE, 1979), the final version was released in 1983. Significant changes were made, which were to reduce the emphasis on the THERP approach and to incorporate two sets of time reliability curves (TRCs), one for screening and the other for final evaluation purposes. Swain's use of the TRC was an attempt to respond to the criticism of THERP in that it did not cover cognitive actions. This issue was raised during the IEEE conference review. The Swain TRCs were derived from expert judgment considerations and, it is believed, were to be guided by the work undertaken by General Physics and Oak Ridge personnel in recording operator responses to simulated accidents (Kozinsky et al., 1983). The industry owed a lot to the work sponsored by the USNRC and organized by General Physics (GP) and Oak Ridge National Laboratory (ORNL) for a number of different reasons.

The handbook developed by Swain and Guttmann was a very complete answer to the issue of the determination of human error probabilities. The handbook addressed the problem of shifting from general error probabilities associated with given tasks to determining the impact of different situations or environments for the same task. The approach was to use modifiers called performance shaping factors (PSFs). The handbook accounted for interactions or dependencies between persons. It also introduced the idea of errors of commission and omission and differentiated between them. The handbook was a complete treatment of HRA and a significant step forward in the technology at that time. Many of the concepts and ideas contained in the handbook are still debated, and we are trying to determine if they are supported by observations of human performance. This goes to underscore the difficulty of human error probability estimation and proving one concept over another.

TRCs were important before GP/ORNL released the results from early simulator sessions with operators responding to accidents. One version was based upon studies undertaken at General Atomics on manual control of high-temperature gas-cooled reactors (HTGR) following full and partial failure of the automatic control systems. The TRC derived from these studies was incorporated into the HTGR PRA, which was called the Accident Initiation and Progression Analysis (AIPA) (Fleming et al., 1975). The data in the later GP/ORNL studies took the form of what was called nonresponse TRCs. They dealt only with successful actions of the operators and not all of the data collected, although the reports contained operator failure data. The GP/ORNL studies were intended to provide information to be used to determine the allocation between manual and automatic actuated controls to perform safety functions and were incorporated into an American Nuclear Society (ANS) standard (ANS 58.8). The ANS standard was an attempt to answer the following: When was it defendable to let the operators take safety-related actions, and when was it necessary to ensure that the action was automated (with sufficient reliability)? The work was funded by the USNRC to support development of an ANS standard 58.8 on this topic (ANS, 1994). Interestingly, the release of the GP/ORNL documents, of which there were a number, lead to an interest in using this data for HRA purposes under the supposition that TRCs reflected the cognitive aspects of human performance.

A number of researchers drew upon these TRC studies, namely, Hall, Fragola, and Wreathall (1982); Hannaman, Spurgin, and Lukic (1984); and Dougherty and Fragola (1988). In addition to the use of time as the principle feature in the determination of human reliability probability, the researchers introduced other aspects, such as using concepts of skill-, rule-, and knowledge-based behavior; burden; stress; and mean task time to shape TRCs. Of these TRCs, the one study that seemed to have the biggest impact was the human cognitive reliability (HCR) model developed for the Electric Power Research Institute (EPRI) (Hannaman, Spurgin, and Lukic, 1984). Probably the reason for this was the role of EPRI in the nuclear industry at the time. The HCR model has been used all over the world by a number of organizations, including such diverse countries as Germany and China, for PSA studies. The EPRI manager of the HCR project felt that the method was useful, but the basic set of assumptions used to develop the model needed to be substantiated. As a result, ERPI funded the operator reliability experiments (Spurgin et al., 1990a) that were supported by six nuclear utilities. These utilities represented users operating both boiling water reactors (BWRs) and pressurized water reactors (PWRs). Many simulator sessions were held at each utility, involving all of their active crews. Each crew was subjected to three to five different complex accident scenarios. Some discussion of the scenarios and the experimental design are carried out in Chapter 12.

The study collected a lot of data, both time and observer data, on the performance of the crews. The data show that the HCR model was not fully justified but did lead to the general characteristic TRCs for both PWRs and BWRs that are of use for training purposes but, in the opinion of this author, do not support the concept of the TRC for reliability purposes. This aspect will be further commented on along with other issues later in this book.

Some other early HRA models are HEART (Williams, 1988) and SLIM (Embrey et al., 1984). These two models were developed in the United Kingdom by different researchers. HEART was developed by Williams and stemmed from his extensive human factors experience in different fields. Embrey and his associates developed the SLIM approach under contract for the USNRC. These two methods have had an impact on HRA analyses in different countries. The HEART method has been extensively used in PRAs in the United Kingdom and continues to be so, although a later version following the same style, called nuclear action reliability assessment (NARA) (Kirwan et al., 2008), has been developed to shore up some of the limitations seen in HEART. A version of SLIM, called failure likelihood index method (FLIM), was used by one of the foremost PRA consulting companies in the United States for some time. The SLIM/FLIM approach uses anchor values to define the range of likely HEPs. Interestingly, the HCR model has been used to provide an anchor value or limit in some Pickard, Lowe, and Garrick (PLG) studies. In the sections to follow, the basic characteristics of the various commonly used HRA models will be commented on. The characteristics of a number of later HRA developments will also be covered. The pros and cons of the various methods are discussed at length in Chapter 7.

It should be mentioned that there was an earlier phase in the development of HRA methods and models in which a number of models or submodels were developed to cover specific effects that investigators believed were missing from the group of HRA models available at that time. One such technique was the confusion matrix method to depict the problems operators might have with discriminating between a small break loss of coolant accident (SBLOCA) and a steam generator tube rupture (SGTR) (Potash et al., 1981).

5.2 HRA MODELS: CHARACTERIZATION

As mentioned by Lyons earlier, a wide range of various HRA models or techniques are available, each with their own characteristics. However, this book is not intended to cover all of the possible approaches. A number of approaches have been singled out based upon their use. It was decided to discuss the following: task-related (discrete nodal) models, task-related (grouped action) models, time reliability models, and context-related models. A separate group of applications use only expert judgment. The problem with this approach is that there is no underlying model or philosophy. Figure 5.1 depicts a set of models that are dominated by one or the other of these characteristics.

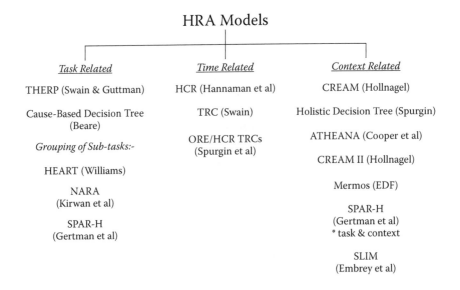

FIGURE 5.1 Various human reliability assessment models grouped by characteristics.

It is possible that the developers of these models and methods can see something different in their models than the author does. Having some identifiable attributes enables one to examine and discuss the basis for the model in a constructive manner. Maybe all are like the blind men trying to determine the characteristic of an elephant (see poem at the front of the book). Perhaps no one approach will ever answer all of the separate issues that can arise in human reliability. Perhaps THERP will always be accepted for task-related situations. However, it is doubted that the database given by Swain will continue be used. Remember all of the changes that have occurred in the nuclear power plant world since 1979, for example, symptom-based emergency operating procedures, intense use of human factors principles for all displayed functions, improved crew training, use of simulators for training, and improved displays and alarms. Also, maybe one can devise methods to test each of the modeling characteristics and see whether they are reasonably correct or not. The USNRC funded an industry-wide study based on trying to test various HRA models using the Halden simulator. It will be interesting to see what comes out of this study (Julius et al., 2008).

A number of models have been selected to discuss, along with their characteristics, and there are other HRA models and methods that also could have been included in this discussion.

The objective of this chapter is to cover a number of the various HRA methods and models. However, the utility of a particular approach is a function of a number of components, not just the absolute value of a method. For example, if one is undertaking a completely new application, appropriate data that could be used may be limited. And in this situation, one may have to select a method using generic data. Equally, if one is selecting a method to embody many years of operational data, then one is unlikely to select a method based upon generic data. For cases where data and experience are available, there are a number of approaches that could be used, including expert judgment. When using just expert judgment, one needs to be careful in the organization and use of the experts and in their selection. The experts selected need to truly be experts in the topic.

The three main HRA model types were indicated previously—namely, task related, time related, and context related. Each type and how these models are implemented are discussed below. In Chapter 7, a discussion of their strengths and weaknesses is presented. Before starting, it is important to discuss an important part of HRA, which is the use of expert judgment. Expert judgment is an integral part of HRA, and it is difficult to proceed in the process of human error quantification without some recourse to its use. Expert judgment is not a model but a process of capturing information about human actions based upon the use of the knowledge and understanding of persons who are either directly involved or observe the actions of others in the operation of a plant. Expert judgment has been used to provide HEPs for given actions or estimates of the impact of subtasks or environments on the resulting HEPs. So the correct use of expert judgment is critical to the HRA process, and to the determination of risk or safety of a plant. In each of the model characteristics, the expert judgment needed to complete the estimation of the HEP values will be pointed out.

5.3 DESCRIPTIONS: MODELS WITHIN THREE GROUPS

5.3.1 TASK-RELATED HRA MODELS

It is proposed that there are two groups of task-related HRA models. The first of these is THERP, and a derivative of THERP is called the cause-based decision tree (CBDT) approach. The second group of task-based HRA methods includes the HEART, the NARA, and the standardized plant analysis risk–human reliability (SPAR-H). SPAR-H appears to fall into both the task group (decision and action tasks) and the context group because of the strong contextual influence of PSFs involved in deriving the crew HEP.

5.3.1.1 THERP

The THERP approach is based on the results of a task analysis, which breaks a task into a number of subtasks. Swain then makes this subtask array into an assembly of discrete HRA subtasks, forming an HRA event tree. To quantify this HRA event tree, Swain then says that one should select the appropriate human error probabilities to match the subtasks in the HRA event tree.

Swain based his model structure approach upon a human factors tool—the task analysis. In the task analysis process, any task is broken down into a number of elements or subtasks. The approach taken by Swain is to allocate an HEP value for each element or subtask based on identifying the subtask with a description in a series of look-up tables. Associated with each description is an HRA estimate (Table 20 in the handbook by Swain, 1983). Each of the subtasks is then depicted in a human reliability event tree. The total of the HEPs in the tree are summed to give an overall HEP. To account for the human capability to correct an error, Swain also introduced the possibility of an operator recovering from an error into the event tree by means of a return path that diminishes the effective failure probability. Figure 5.2 depicts such an HRA event tree. Swain's approach is very much an engineering approach to a human error modeling problem.

5.3.1.1.1 Cause-Based Decision Tree (CBDT)

A task that Beare was assigned as part of the operator reliability experiments (OREs) team was to try to incorporate the data collected during the ORE set of operator experiments into something that could be used in HRA. He was not able to directly use the data and experience from ORE, but from his experience in the GP/ORNL and ORE simulator projects and with using THERP, he saw the utility of using a decision tree approach to try to incorporate the ideas stemming from ORE and THERP.

His approach was to construct a set of decision trees based upon the Swain tables and other considerations, including some derived from the ORE experience. Instead of the human reliability analyst having to identify the key subtasks associated with the overall job and then going to the tables, the HRA analyst could examine each of the eight trees (grouped into two higher-level effects) and see what the pathway through each tree that corresponded to yield an HEP based upon the analyst's view of the set of influences. The HRA analyst would then sum the outputs from each individual tree.

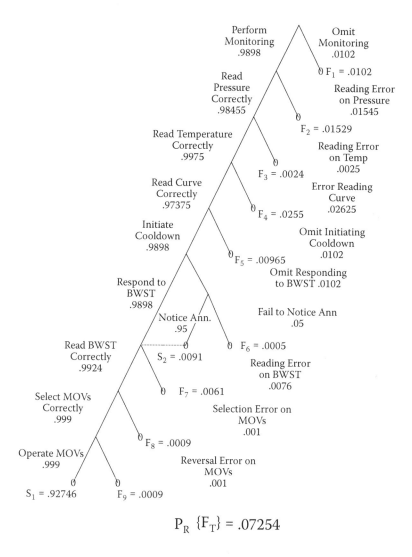

Perform
Monitoring
.9898

Omit
Monitoring
.0102

$F_1 = .0102$

Read
Pressure
Correctly
.98455

Reading Error
on Pressure
.01545

$F_2 = .01529$

Read Temperature
Correctly
.9975

Reading Error
on Temp
.0025

$F_3 = .0024$

Read Curve
Correctly
.97375

Error Reading
Curve
.02625

$F_4 = .0255$

Initiate
Cooldown
.9898

Omit Initiating
Cooldown
.0102

$F_5 = .00965$

Omit Responding
to BWST .0102

Respond to
BWST
.9898

Notice Ann.
.95

Fail to Notice Ann
.05

Read BWST
Correctly
.9924

$S_2 = .0091$

$F_6 = .0005$

Reading Error
on BWST
.0076

Select MOVs
Correctly
.999

$F_7 = .0061$

Selection Error on
MOVs
.001

Operate MOVs
.999

$F_8 = .0009$

Reversal Error on
MOVs
.001

$S_1 = .92746$ $F_9 = .0009$

$$P_R \{F_T\} = .07254$$

FIGURE 5.2   A representation of a human reliability assessment event tree.

Thus in the CBDT approach, a set of subtasks are placed in a decision tree or event tree format. In the Swain approach, the HRA event tree is like a single-sided herringbone structure, whereas the CBDT structure is more like a conventional event tree, except there are a number of such trees. In both cases, the effective overall HEP is arrived at by summing the individual HEP values from the bones of the herring or the output from each separate decision tree. CBDT formulation is one of a set of HRA methods included within the current EPRI HRA calculator (Julius et al., 2005).

There were some differences between CBDT and THERP. CBDT is more than an evolution in the presentation of Swain data. Beare had mixed both PSF and HEP numbers in his formulation, which is a little hard to justify. He also drew upon data and insights from the ORE study, general practices in HRA, along with modifications incorporated from the Accident Sequence Evaluation Program (ASEP) 1987 report. It is believed that EPRI has made changes to the Beare CBDT formulation as incorporated in the latest versions of the EPRI HRA calculator.

Beare identified two main human error failure modes: "Failure of the Plant Information–Operator Interface" and "Failure of the Procedure–Crew Interface." For each of these error failure modes, he constructed four failure mechanisms. For each of these mechanisms, he constructed a decision tree. One such tree is depicted in Figure 5.3. This tree depicts predictions for HEPs for a case when information used by the operator is misleading. Beare identified this as $p_i d$. He provided descriptions for each of the headings (e.g., heading #1 is "Are all cues are as they are stated in the procedure, if yes then proceed up, if not then proceed down"). The same process goes for the rest of the branches. Beare defined the specific areas that can lead to failure probabilities.

He stated that there are four mechanisms associated with the Plant Information–Operator Interface: (1) data not physically available for the main control room (MCR) operators, (2) data available but not attended to, (3) data available but misread or miscommunicated, and (4) available information is misleading.

Further, Beare identified four mechanisms associated with the failure of the Procedure–Crew interface. These are as follows: (1) the relevant step in the procedure is skipped, (2) an error is made in interpreting the instructions, (3) an error is made in interpreting the diagnostic logic (basically a subset of 2), and (4) the crew deliberately violates the procedure.

His original report contained guidance on a number of issues, including the possibility the crew would recover from an error. He also covered cases where he had moved away from Swain's HEP values and why. Such situations were discussed that have been seen in practice, such as the TMI Unit 2 accident in which recovery was due to the arrival of an outside person during the accident progression. So this method, although heavily dependent on Swain's work, is more than the reorganization of Swain's "data." To use the approach, one needs to be quite experienced in the application of HRA concepts and have some experience of how control room

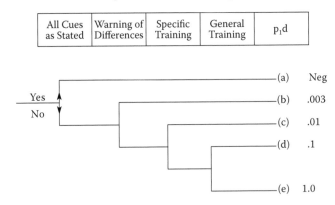

FIGURE 5.3 Decision tree representation of $p_i d$, information misleading.

crews operate (i.e., best seen during simulator sessions). In other words, the method cannot be blindly applied and then expected to produce meaningful results. In fact, utility HRA personnel who have used the CBDT approach have made changes to the branch end state values to more fit their experience (see Kohlhepp, 2005).

Now you may ask, how does one differentiate between tasks being carried out in this plant versus that plant? Well, Swain introduced the concept of the performance shaping factor (PSF) to achieve this. It is not sure that Swain came up with the idea, but he was the first to incorporate such an idea into HRA as far as is known. The concept of the PSF is a key element in a number of HRA methods. So with its use, one can model various tasks at different plants and then modify the HEP values accordingly. The question to be asked then becomes, is it correct? As discussed later, this PSF process is used in a number of HRA methods in one form or another. The idea more simply put is you can capture the essence of a situation in which a task is being performed, purely by quantifying a task and subsequently adjusting the basic reliability by the use of PSFs. Of course, this makes the topic of human reliability very compatible with engineering equipment reliability, so you can have tables to determine HEPs and you do not need to understand the real impact of the context in which the human is working. But does this explain why a similar task in one station has a high failure rate and in another plant a higher success rate? It would appear that, based upon THERP, they should be the same within a small error bound. What is different? Well, the whole context is different, but to the outside analyst, there appears to be very little difference. A small change in the control board displays, or much less emphasis in training for a given accident and the attitude of management relative to a given accident's importance, should lead to a different assessment at one plant versus another.

5.3.1.2   The Second Group of Task-Related HRA Methods

One of the problems with the Swain method felt by a number of users was that the combination of the subtasks and the quantification process was too limiting. Although there are a number of subtasks, even a combination of identified subtasks may not cover nuclear operations very well.

5.3.1.2.1   *HEART*

One of the first to modify the tasks to address this problem was Williams, who thought one should quantify the whole task rather than build the complete task from the sum of subtasks. So in a way, Williams was thinking more holistically about the implementation of HRA to the modeling of operations. His HRA development was called HEART (Williams, 1988). He brought to the process a large amount of experience to human factor effects from different industries. He also developed a different approach to compensate for PSF effects, and defined a number of different PSFs, related to his experience, and a method to modify the basic task with each of these PSFs. A range for each PSF was defined. The key elements of HEART are a listing of a number of tasks in tabular form along with an associated mean HEP and a range of values to cover uncertainties in the estimates. The method also covers a number of weighting factors, which are introduced to cover the potential influences of a range of PSFs. The weighting factors are introduced using the following formulation:

$$WF_i = [(EPC_n - 1) \times APOA_n - 1.0] \qquad (5.1)$$

$$HEP = GTT_1 \times WF_1 \times WF_2 \times WF_3 \times ...etc. \qquad (5.2)$$

where GTT_1 is the central value of the HEP task (distribution) associated with task 1, GTT stands for generic task type, EPC_n is the error producing condition for the nth condition, $APOA_n$ is the assessed proportion of "affect" for that condition, and WF_i is the weighting effect for the ith effect.

It is useful to discuss the basic items that Williams generated. He defined a limited set of tasks to describe activities within a nuclear plant for use in the PRA and called these generic tasks (GTTs). It was his intent to cover a sufficient number of such tasks so that the HRA analyst could pick an appropriate one for his analysis. He advanced the idea of an error producing condition, which if this was present, it would change the HEP distribution, making the error more or less likely. This operates like a PSF. However, the error producing condition can be modified according to the actual situation, and this is the APOP or assessed proportion of the error producing conditions (EPCs). His concept then adjudges all EPCs relative to each other, so there is a relative level of importance between EPCs. However, in a particular situation, a given EPC may be more or less important; therefore the value of the EPC is increased or reduced by the application of the APOP.

Another point that Williams makes is that in executing a task, the crew may be subject to a number of EPCs, not just one. He lists the range of these APOP factors, so the relative importance of a given EPC with respect to other EPCs does not change, but its effect in a scenario is varied by means of the APOP. It is interesting that he should do this. It suggests that the relative importance of EPCs is fixed in some time frame and therefore suggests that this relative importance can be established by theory or experiment. Later, the same kind of question can be asked of NARA.

If the analyst decides that there are a number of EPCs at work, these are accounted for by Equation 5.1. In other words, each EPC_i is modified by the use of the $APOP_i$, resulting in a weighting factor W_i. In turn, each of the Wjs multiply the GTT to produce the calculated HEP_i, as given by Equation 5.2.

Although HEART lists various tasks and gives descriptions of these tasks, the HRA analyst has to select the nearest HEART task to the actual task to be modeled. HEART also gives a value for each EPC and a range of values for each APOA and describes them in the context of the shaping factor; the selection of the precise value for each factor is up to the analyst. HEART can be used to cover multiple tasks within the PRA structure.

If the analyst wishes to analyze an accident with multiple tasks, which can coexist, then there is a possibility of interaction between the various HRA tasks. Because there are few tasks identified by Williams, it is not sure whether he was moving toward multiple tasks to describe the actions of the station personnel or toward a single comprehensive task. The existence of multiple human tasks depends on the analyst and an understanding of the relationship of a HEART task in the context of the accident. If the analyst chooses to model the human action at a top level, then there will be no human dependency effect. However, if the analyst chooses to model separately the actions of the humans, then there is a dependency effect. Therefore for this case, there is a need within HEART to consider dependency effects. Williams referred to the dependency formula given by Swain. This is pretty much the case for a number of HRA developments. However, it is interesting to note that a number of plant HRA analysts developed their own dependency compensation formulas; see Chapter 6 on HRA tools (dependencies).

5.3.1.2.2 NARA

NARA is a further development of the HEART process to cover more tasks, and these are thought to be defined better relative to nuclear power plant operations. Despite the fact that HEART has been used in the British nuclear energy field for a number of PRAs, there have been questions raised about the HRA probability numbers based on HEART and the defense of these numbers. As an aside, one of the early development versions of ATHENA (Cooper et al., 1996) used the HEP values given in HEART; the latest version of ATHENA (Forester et al., 2007) uses expert judgment to derive HEPs.

To meet the criticism of the HEART database and its justification, the developers of the NARA model have used a more developed database called CORE-DATA (Gibson et al., 1999), rather than HEART's associated database that was considered not to be completely defendable. Much of the structural elements of NARA are close to the HEART formulation. The differences lie in three areas—the use of CORE-DATA, the substitution of the HEART tasks with a set of NARA tasks, and the incorporation of a human performance limit value (HPLV) when multiple HEPs occur together.

On this later point, there is a feeling within the HRA community that no matter what steps are taken to enhance the reliability of human operations, it is impossible to achieve HEPs less than, say, 1.0E-4 or some such number. NARA introduces a limit process to take care of these considerations. It should be pointed out that Wreathall (Hall et al., 1982) produced a TRC that addressed the same concern in 1982. The time reliability curve he produced at that time had two limits of 1.0E-04 and 1.0E-05 to address the selfsame concern. Some of us think that the number might be nearer 1.0E-03. Even the early GA HRA TRC had a limit of 1.0E-03.

NARA identified four groups of GTTs: A, task execution; B, ensuring correct plant status and availability of plant resources; C, alarm/indication response; and D, communication. One can see that NARA developers tried to define often simple tasks related to more complex tasks seen during crew responses to accidents, for example, one might see the crew tasks responding to a number of NARA tasks. The NARA type A class calls for the crew to take action. This act is usually the last action in the crew response to an accident. The first action may be in response to an alarm (e.g., if a type C, then the crew would discuss the situation [type D], communication). Before the crew takes an action, they check the availability of systems and components type B. In a way, the NARA model lies between HEART and THERP methods. The methods are focused on the task for all three models, it is only how the

TABLE 5.1

Generic Task Type (GTT) Information for Two GTTs

Nuclear Action Reliability Assessment GTT ID	GTT Description	GTT Human Error Probability
A1	Carry out simple manual action with feedback. Skill based and therefore not necessarily procedure	0.006
A2	Start or configure a system from MCR following procedures, with feedback	0.001

task is broken down. The task to be executed by the crew still determines the basic HEP, and this is central to this basic approach.

For each GTT, there is an HEP value (e.g., see Table 5.1 for two such GTTs).

NARA defines some 18 EPCs and gives anchor values for each EPC and an explanation of the anchor values, along with how to select APOP values. The descriptive material associated with each EPC is extensive in the NARA documentation, so Table 5.2 indicates the description of an EPC and its associated APOA, including the result of applying the equations similar to Equation 5.1.

As stated above, the assessed HEP would be arrived at by multiplying the GTT HEP value by a set of weighting factors (W_j). The selection of the GTT and associated EPCs would depend on the capabilities of the user to evaluate the HRA task and then make the selection of the closest description that fit the task, the type of error producing conditions associated with the task, and the corresponding weighting factors considering the task.

One issue of importance is related to whether one is trying to match the average crew response or that of an individual crew. There seems to be some confusion in the application of HRA technology to a PRA. Due to the randomness of an initiating event, the likelihood of a specific crew being present is low; therefore one should be dealing with an average crew. An average crew is affected in a systematic way rather than a specific way. Therefore the analyst has to consider effects that are systematic (i.e., that can affect all crews rather than a given crew). The approach deals with an HEP distribution, so this distribution should cover the variability of crews. In practice, the variability of individuals is compensated for by design in the use of teams or

TABLE 5.2

Example of the Nuclear Action Reliability Assessment (NARA) Error Producing Condition (EPC) Weighting Factors

EPC	Description	EPC Factor (f)*	APOA(p) Estimated	Assessed Factor (f-1.0)xp-1
EPC03	Time pressure	11	1.0	9.00

* Obtained from Kirwan, 2008, Table 2.

crews rather than individuals. The influence of outlier persons is mitigated by both the selection of operators and by the use of crews.

5.3.2 TIME-RELATED HRA MODELS

The concept behind the time reliability curve (TRC) HRA approach is that a crew will eventually respond to an accident given enough time, so the estimated HEP decreases depending on the time available before an accident reaches an irreversible point. Some will argue with this point of view—for example, the crew during the TMI Unit 2 accident was not aware of the significance of the accident until an outside person pointed this out to them some time into the accident.

5.3.2.1 Swain TRC

Swain's TRC was added to the handbook document after there was criticism of the handbook failure to account for cognitive effects, as depicted in Figure 5.4. The Swain TRC uses three curves to predict the HEP median value and distribution (5% and 95%) as a function of time. The estimated time window intersects with the TRC and yields an estimated HEP and range (see Figure 5.4). As mentioned in Section 5.3.1.1, Swain also suggested the use of PSFs to modify his TRCs. Swain's TRC is different than other TRCs in that the curve has a tail and the curve represents nonsuccess rather than nonresponse curves given in the GP/ORNL series of reports (Kozinsky et al., 1983). The later simulator data collection projects (Bareith, 1996) focused more on nonsuccess rather than nonresponse, so that failures and later

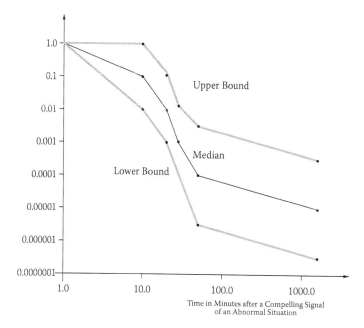

FIGURE 5.4 Swain's time reliability curve.

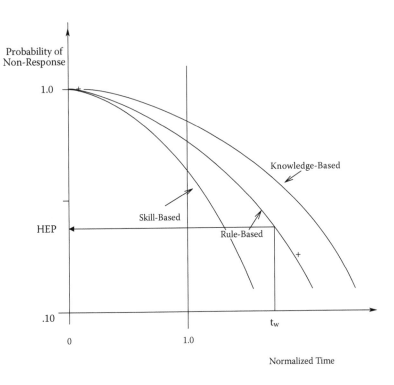

FIGURE 5.5 Human cognitive reliability, time reliability curve developed by Electric Power Research Institute.

recoveries could be captured rather than focusing only on successful operations. People forget the objective of the GP/ORNL work was in times when the operators were successful, whereas the later simulator data collection projects were more interested in the totality of the response, including failures. This data could be used for both training and HRA purposes.

As a result of the developing interest in TRCs stemming from the simulator data collection work sponsored by the USNRC by Oak Ridge Labs and General Physics, EPRI sponsored the development of the HCR approach (see Figure 5.5). The HCR method was an approach based on the earlier simulator studies together with the concept of scaling a basic TRC by using a mean task time and modified by whether the task was considered to be skill, rule, or knowledge based (SRK) (Rasmussen, 1979). The HEP was estimated by examining the intersection of a time window with the appropriate SRK curve. Because the basic HCR curves were normalized curves, the curves had to be redrawn using either the estimated task time or a modified time window given by $t_w = T_w/T_m$, where T_w is the time window and T_m is the mean time for the execution of the task. Figure 5.5 shows the normalized HCR curve showing the three SRK characteristic curves. The figure also shows the intersection of the time window with a rule-based curve. The HEP corresponding to the intersection is read off on the probability of nonresponse axis. The time window was estimated by transient analysis considerations and was the time that some selected plant variable reached an unacceptable value, such as the onset of core damage. The selection of the conditions for the time window is normally made by the PRA team.

Later, EPRI sponsored a large simulator data collection project (ORE) (Spurgin et al., 1990a) with the aim of support or otherwise of the underlying hypotheses of HCR. The result of the study was that not all of the hypotheses were confirmed. However, the project achieved a number of objectives, one of which was to generate insights into how the crews performed and give rise to the development of the holistic decision tree (HDT) method based upon context driven rather than task driven, see later. Also, it became quite clear that the TRC is just an expression of the variability of persons. As a tool, the TRC is quite useful in training to see if MCR crews can reach needed responses within the time limits of the task and with sufficient confidence.

For any of the TRCs to be used, one needs to define a time window for an accident involving the task. The interaction of the time window with the TRC defines the HEP value. In practice, often the HEP value can be very low, when based on this approach <1.0 E-8. As mentioned above, several researchers have suggested a lower limit of 1.0E-4 or 1.0E-5 in cases where the indicated number is very low.

In the HCR approach, the TRC is based upon "normalized time," and the curve needs to be modified by multiplying the normalized time by the mean task time ($T_{1/2}$), then the same rules are applied as for a normal TRC. The analyst has to select whether the task is S, R, or K and then use the corresponding normalized TRC. EPRI produced a set of TRCs for PWR and BWR plants for various procedure rules. These could be used in a similar way to the HCR curves.

5.3.3 CONTEXT-RELATED HRA MODELS

Context-related HRA methods are quite different than the task-related or TRC-related HRA methods. In the other two methods, the task or time element are the important items in predicting the HEP value. For the context-related methods, the context under which the action takes place is important, not the task or time. It should be stated that the context is directly related to the task. An accident calls for a response by the crews to terminate or mitigate the accident. The task to be performed is the result of the accident and the procedures selected by the crews. The selection of the procedures depends on the training of the crews, the information used by the crews, and their interpretation of that information (i.e., the crew's experience in dealing with specific events). The accident progression depends to some extent on the communications between the crew members. Thus the quality of the training, man–machine interface (MMI), procedures, and communication standards bears heavily on the ability of the crew to meet its target of controlling the accident and its progression. The task is built into the fields of operation and defined by the procedures being followed by the crew. The quality of the procedures is important in defining the task. The concept of the task is really an artifact of the plant designer or operations designer. The operators are led by the indications and their timing on the MCR as a result of the accident to select the appropriate set of procedures. Discussions among the crew help to clarify the process. The crew then responds to various steps indicated by the procedures that were selected. The operators' actions are what defines the task, but after the fact. The trajectory taken by the crew in responding to an accident may differ from that determined by analysts either before or after an accident, but still may be successful judged by whether the correct end state is met.

The HEP is determined by each of the influential context elements. Clearly, some of these are the quality of training of the crew, the quality of the procedures, the quality of the MMI, the quality of the communications standards, and so forth. The important context elements depend on the situation being considered. The quality of any context element can vary from poor to good to excellent to superb. Over a range of accidents, the context can vary depending on the amount of attention given to each context element by the designer or plant management (e.g., the quality of the MMI in one accident may be good and in another accident poor).

5.3.3.1 Holistic Decision Tree Method

One context method was developed by the author after being deeply involved in observing control room crews responding to accident scenarios on simulators during the EPRI-funded operator reliability experiments (ORE), in the period 1986 to 1990. The method was called the holistic decision tree (HDT) method. The ORE results indicated that the operators were more influenced by context during the execution of actions to accidents rather than nebulous task concept. An event type of structure was tried to see if the results of the observations could be captured.

Because of the involvement of the project team, these concepts were passed to GP and Beare along with all of the ORE results and observations for them to try to formulate a useful HRA method. The result of their work was the CBDT approach mentioned earlier.

Subsequently, the original decision tree method was carried forward and implemented in the early EPRI HRA calculator in the form of trees for use in the determination of latent or preinitiator errors. This work was funded as part of the HRA calculator development under a joint project between EPRI and Texas Utilities on a HRA calculator tool for the Comanche Peak PRA to provide a method for recording HEP ranges and looking at sensitivity calculations. The HRA calculator was meant to be a support tool for the HRA analyst and a means of keeping records.

The HDT combines a tree structure with anchor values to determine the end-state HEPs for a particular accident scenario; this approach has some connection to the SLIM HRA approach. The method is directed toward a holistic approach to the estimation of HEP values for the MCR crew, although the initial attempts at modeling HEPs was aimed toward estimating latent types of failure, such as those due to maintenance and test operations. This approach was incorporated into the first EPRI HRA calculator produced as a joint project between Texas Utilities and EPRI in 1992 (Moeini, Spurgin, and Singh, 1994, Part 2).

The next phase in the development of HDT was as a result of collaboration with a Hungarian research group (VEIKI) in trying to capture results and experience from their simulator sessions at Paks NPP, the results of which were to be used in the Paks PSA. The first comprehensive HDT HRA approach was developed based on the results of the simulator sessions carried out with the full complement of Paks NPP control room crews. In addition, Paks technical staff developed a data collection tool to be used with the simulator, and it was data from the simulator along with observer comments that were used to support the HDT model (Bareith, 1996).

A further improvement of the HDT approach was made in conjunction with some Russian PSA experts on the SwissRus project (Spurgin, 2000). The HDT approach has been applied in a number of Russian PSAs. An advocate for the use of the method in Russia was Irena Kozmina of the Russian team working on the Novovoronezh 1000 MWe NPP.

Because the HDT method is not too well known, it was decided to give a comprehensive description of the method and how it is applied in Chapter 10. A brief overview is presented below. Please note that the influence factors (IFs) are similar to PSFs, and weighting factors associated with HEART and NARA are similar to quality factors (QFs).

Figure 5.6 gives a partial implementation of a context HRA model for the International Space Station as a Microsoft Excel® program. The top line covers

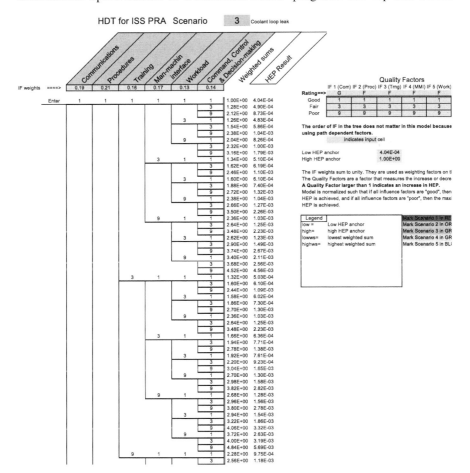

FIGURE 5.6 Holistic decision tree human reliability assessment model implementation based on Microsoft-Excel®.

various influence factors (IFs), and each has a range of different quality factors (QFs), shown in table to the right of the main figure. In practice, each of the QFs are described in detail and based upon the associated technology (e.g., the MMI uses human factors descriptions of what may be considered to be good to poor). It should be noted that IFs are used here, but other HRA developers may use PSFs. However, PSFs are associated with those factors that are used to modify predefined HEPs to account for variations in context associated with different plants, as opposed to using the factors to define the human error. This difference comes about from a different philosophy on the causes of human error. Some of the methods like SPAR-H and CREAM have a strong emphasis on PSFs or equivalent and lesser emphasis on the importance of the task. Philosophically, these methods should be task oriented, but because they have strong reliance on PSFs, they are grouped in with the context models. SLIM was the first context model and used a series of PSFs to determine an HEP relative to a set of anchor values, in other words the approach used by Embrey was to set up a relative measure and then relate this to two anchor values so the actual HEP fell within these anchors.

5.3.3.2   CREAM II

CREAM II (Hollnagel, 1998) is called by its developer extended CREAM. CREAM II is based upon two concepts: a generalized cognitive failure function (CFF), which is divided into four groups: observational errors, interpretational errors, planning errors, and execution errors (see Table 5.3). There are 13 errors and a functional modifier called common performance conditions (CPCs) of which there are nine. These CPCs determine the context under which the crew or personnel operate. The CPCs go from things like the adequacy of organization to crew collaboration quality.

Rather than task identification, the method concentrates upon cognitive characteristics associated with a task. CREAM II is not included in the task section because of its emphasis on context (i.e., the use of CPCs). The method is more associated with context-related methods like HDT (Spurgin, 1999) than, say, NARA, but of course, others may choose other groupings.

Clearly, this is a second-generation model in the development of HRA in that it is moving to more of a phenomena approach (i.e., cognitive effect and context impact). It is based upon a more psychological view of HRA and is context driven, but the context is considered to affect some aspect of the cognitive processing of the operators, and this leads to functional failure, which in turn leads to an error. The general concepts are very much in line with current thinking about how errors are caused. In other words, the human is set up to succeed or fail depending on the context under which he or she is operating. It starts with the idea that humans do not deliberately set out to produce errors. Of course, there may be work conditions that lead to a worker taking action to cause problems (e.g., see Chapter 8 and the case of the Bhopal accident).

Hollnagel considers that there are potentially 13 cognitive errors associated with a task. These are three observational errors, three interpretational errors, two planning errors, and five execution errors. A detailed description of each of these errors is given in Table 5.3. These are quite high-level descriptions of potential errors. The analyst is faced with the job of trying to assess in an accident sequence how the error

TABLE 5.3

Listing of CFC Numbers and Definitions

Cognitive Function	Potential	Cognitive Function Failures
Observation errors	O1	Observation of wrong object. A response is given to wrong stimulation or event
	O2	Wrong identification made, due to, for example, a mistaken cue or partial identification
	O3	Observation not made (i.e., omission), overlooking a signal or measurement
Interpretation errors	I1	Faulty diagnosis, either a wrong diagnosis or an incomplete diagnosis
	I2	Decision error, either not making a decision or making a wrong or incomplete decision
	I3	Delayed interpretation (i.e., not made in time)
Planning errors	P1	Priority error, as in selecting the wrong goal (intention)
	P2	Inadequate plan formulated, when the plan is either incomplete or directly wrong
Execution errors	E1	Execution of wrong type performed, with regard to force, distance, speed, or direction
	E2	Action performed at wrong time, either too early or too late
	E3	Action on wrong object (neighbor, similar, or unrelated)
	E4	Action performed out of sequence, such as repetitions, jumps, and reversals
	E5	Action missed, not performed (i.e., omission), including the omission of last actions in a series (undershoot)

is caused (i.e., is it because of a man–system interface issue, procedure deficiency, or something else).

Hollnagel (CREAM) then selected data from a variety of sources, such as Beare et al. (1983), THERP (Swain and Guttman, 1983), HEART (Williams, 1988), and Gertman and Blackman (1994) to populate a tabulation of error probabilities corresponding to the above errors; a typical example is O1, the base value is 1.0E-3, the lower bound is 3.0E-4 (5%), and the upper bound is 3.0E-3 (95%).

Hollnagel identifies performance modifiers as CPCs. The CREAM user has to evaluate a situation and then select the CFF and also determine which CPCs are involved and to what degree. An example might be if an action is performed out of sequence (CFF-E4), the CPCs might be the working conditions, and they are considered helpful. The other CPC is the adequacy of training and preparation, and this is considered adequate.

The combination of these are CFF central value 3.0E-3, CPC2 = 0.8, and CPC4 = 0.8. CPC2 relates to working conditions, and CPC4 relates to the availability of procedures or plans. The weighting factors mean here that they are considered advantageous. Therefore the resulting mean probability equals 3.0E-3 × 0.8 × 0.8 = 1.92E-3 ~ 2.0E-3.

CREAM II is somewhat similar to the high-level concepts of both HEART and NARA, in that they consider a central item or HEP, corresponding to a task or cognitive action, and then change the value by a set of modifiers. In the case of CREAM, these are called CPCs and are weighting factors equivalent to the composite of EPCs and APOPs for HEART and NARA. In the case of CREAM, the modifiers are double sided (i.e., both positive and negative effects are considered). The weighting factors can either decrease an HEP or increase an HEP. So the weighting factors are multipliers, and they are either less than one or greater than one. Performance enhancers are less than one, and those that cause deterioration in performance are greater than one.

For HEART and NARA, only values that increase the error probability are considered. SPAR-H is discussed next, and its PSF weighting will be covered.

5.3.3.3 SPAR-H

SPAR-H (Gertman et al., 2004) was developed by Idaho National Laboratories for the USNRC Office of Regulator Research (RES), Office of Reactor Regulation (RR), and the Reactor Oversight Process (ROP). SPAR stands for standardized plant analysis risk, and HRA, of course, stands for human reliability analysis. The method has been applied in a large number of cases (70 at the time of the INL report). The object of developing SPAR-H was to address the following USNRC-related needs to "account for human errors when: (a) performing safety studies such as probabilistic risk analysis (PRA); (b) helping risk-informed inspection processes; (c) reviewing special issues; and (d) helping risk-informed regulation. HRA has also been used to support in the development of plant-specific PRA HRA models" (Gertman, 2004).

The SPAR-H model builds upon a number of years of experience in the nuclear energy field of the authors, especially in human factors and HRA. The underlying psychological basis for the SPAR-H construct is the informational model of humans. This being the case, they have in mind a diagnosis and action model for crew and personnel responses to accident conditions. They further realize that the responses of persons are affected by the context of the condition under which the persons operate. Their model consists of probabilities associated with diagnosis and action. They take the HEP values as 0.01 and 0.001 for diagnosis and actions. The effective HEP is made up of these elements along with modifiers stemming from the context. The context elements will be discussed later.

As clearly stated in the foreword to the report, the SPAR-H is a simplified approach to HRA, and there are other more comprehensive methods that might be used for more intensive studies (e.g., the USNRC mentions ATHENA, another USNRC project). However, SPAR-H fits well for the use envisioned by the NRC for it. The authors were required to have something easy to use and functional for PRA investigations under the standardized plant review. In order to make the system more workable and give consistent results, the authors selected a series of worksheets to be used for each situation being evaluated. There are sets of worksheets for diagnoses, actions, calculation of base HEPs, and dependencies. Depending on the number of PSFs considered, there are rules that apply for PSFs less than 3. The base HEP equals the diagnosis HEP or action HEP multiplied by weighting factors defined in eight categories within a worksheet. The eight categories are available time, stress/stressors, complexity, experience/training, procedures, fitness for duty, and work processes. For each category, there are a number of judgments that the user of the worksheet has to make. For example, in the case of experience/training, there are four judgment categories: low, nominal, high, and insufficient information. Corresponding to each of these judgment categories is a weighting value. In this case, the corresponding values are 3, 1.0, 0.5, and 0.1. In completing the worksheet, the user is expected to justify the selected category but not the assessed weighting value. Once the worksheets are completed, they are reviewed and signed off.

The final HEP values are arrived at by multiplying the nominal HEP by the weighting factors derived from the tables. This process is carried out for diagnosis and action items, and the overall value is given by the addition of both the diagnosis and action contributions. However, because of concerns about the correct modeling of the contributions, the following rule pertains.

For the case when the number of PSFs, for which the weighting factors is greater than 1.0, is greater than or equal to 3, then the base HEP value is given by the following formula:

$$HEP = NHEP \times PSF_{composite}/[NHEP. (PSF_{composite} - 1) + 1] \qquad (5.3)$$

where HEP is the effective error for either diagnostic or action error (there are separate tables for both); NHEP is the nominal HEP and is either 0.01 or 0.001, corresponding to diagnosis or action; $PSF_{composite}$ is the sum of all PSFs.

SPAR-H adopts the above correction to account for the multiplicative effects of PSFs leading to HEP greater than 1.0.

As in most HRA methods, the effect of dependence should be considered when a number of HRA contributions exist in the plant logic model and appear in the same cut-set. The approach taken in SPAR-H is a modification of the method evolved by Swain. The difference is that SPAR-H uses a defined decision tree formulation with headings generated from their experience. These are crew (same or different), time (close or not close), location (same or different), and cues (additional or no additional). Depending on the pathways, the results are complete, high, moderate, low, or zero dependence. Swain gives equations for all of the first four in the handbook in terms of the initial HEP and weighting factors. The authors of SPAR-H have a dependency condition table, equivalent to a decision tree, in the At-Power worksheet along with the Swain factors.

The other issue that HRA practitioners should consider is uncertainty. The values of HEP calculated are "best" estimate numbers, and PSFs are just weighting values and do not have uncertainty ranges associated with them. SPAR-H authors used a beta distribution to cover the uncertainty distribution.

The NUREG/CR-6883 report on SPAR-H is comprehensive and should be consulted if one is interested in finding out more about the approach and its application.

5.3.3.4 MERMOS

The Methode d'Evaluation de la Realisation des Missions Operateur pour la Surete (MERMOS) is the current HRA used by Electricité de France (the French electricity company). EDF has been involved in PSAs for some time. They published their first PSA report on the three-loop PWRs in 1985. The HRA was based upon a TRC like Swain's but modified to account for experience in carrying out simulator sessions under accident conditions. As a group, the French investigators were more interested in the effects of accidents on the behavioral responses of the crews rather than their actions in time. Consequently, their simulator program consisted of a large number of different scenarios but involved a small number of crews. This was different than the EPRI ORE project, which focused on a small number of scenarios and a large number of crews. In the ORE project, all of the NRC licensed crews from six stations participated in the simulated accident studies.

EDF dropped the "old" HRA approach and has over time developed the MERMOS approach. MERMOS can be considered a second-generation method in that it places more consideration on the context rather than on the task. Although EDF published a number of papers on the topic, it is difficult to completely grasp the MERMOS concepts. The MERMOS authors have a number of ideas and descriptions relative to HRA, and some of their terminology differs from that of other HRA developers. One of the clearest expositions of their ideas is in a paper by Pesme et al. (2007). Three things of note are the idea of the human reliability mission, the consideration of the set of circumstances that can occur as a result of an initiating event, and failure of human or equipment and their interactions. The breakdown of the accident sequence into a number of branches is the human reliability task. Insights into the breakdown are led by EDF's experience stemming from their years of observing simulator sessions at various power plant simulator sites. In addition the simulator experience yields the possibility of actual data, because a large number of EDF's power plants are nearly identical, and these sessions can be used to train experts to make more informed estimates of the crew human error probabilities.

It is difficult to produce a blow-by-blow description of the MERMOS approach without a deep knowledge of the method and its application (e.g., what actual data are available for a given action). Some issues are associated with data, and others are associated with search techniques used to determine the sequence of events and the corresponding stopping rules. It is sure that the simulator sessions may not yield significant statistical operator errors, and this should be similar to the experience gained during the ORE project. It is believed that EDF operator failure rates are probably close to those of the U.S. utility NPPs.

MERMOS states that the probability failure of the human factors mission is

$$P = \text{Sum } [P_{SCI}]_{I=1 \text{ to } N} + P_{residual} \qquad (5.4)$$

The probability P_{SCI} is the composite probability resulting from a given operator behavior associated with a plant feature modified by how the accident affects the situation. Additionally, the MERMOS investigators have included a residual human error probability to ensure the probability does not fall below this figure.

One of the first things that one comes across reading the various MERMOS papers is the term *CICA* (Pesme et al., 2007). A translation of the French might be "configurations/orientations of the system," and this indicates that the combination of operator responses to accidents is covered by a number of CICAs. If one considers the event sequence diagram (ESD), the various elements within the ESD can be represented by CICAs. These are more than just the actions taken by the crew. The MERMOS method is concerned with what is the system, what is the initiating event that causes an accident, and what is the complete mission of the crew and the safety equipment. The approach overall is the same as most PRAs, but the thinking about the combination of things is different. The crew has a number of possibilities in dealing with the impact of the initiating event on the plant, and the crew's responses are going to be affected by the availability of safety and other equipment. All of these things have to be taken into account, and the story behind all of this has to be laid out. Sometimes in U.S. PRA, much of the considerations of what goes on in an accident sequence, including equipment availability, was formulated earlier, and therefore the results of this have been subsumed in later PRA studies.

MERMOS is an attempt to effectively formulate this consideration in a more formal manner. It is believed that EDF was forced to consider anew the process, in part because of the impact of the N4 PWR control room design features that change the relationship between the operator, and in part because of the automated decision making built into the computer-based emergency procedure system.

MERMOS is a significant piece of work that has been undertaken by EDF. The method had to answer questions resulting from a different relationship between man and machine. Some of the processes are understood from the published papers, but to really understand everything, one needs a much better understanding of the parts and the underlying philosophy of the method.

The idea of using careful observations of crews responding to accidents for the purposes of HRA is good. Actual data in terms of the numbers of failures per number of trials is unlikely to occur. However, the steps operators take and their approach to problems can yield insights for both domain and knowledge experts to use to try to estimate human error rates in both similar and dissimilar situations. Building an information base founded upon the observations made of simulated accidents can be a valuable resource for the future. One can use the information to see if the feedback from prior accident responses has been useful and the reliability of crews has increased or otherwise. MERMOS appears to be a useful process, but one is left with a number of questions because of the lack of documentation.

6 HRA Tools and Methods

INTRODUCTION

This chapter contains sections on various tools and methods, including dependencies and errors of commission and omission. In addition, the chapter covers the use of HRA computer-based tools, such as EPRI's HRA calculator. The section on the HRA calculator also covers the history of HRA calculation support tools. Philosophical differences exist between the developers of these tools, and this is mentioned in the corresponding section.

6.1 HUMAN RELIABILITY ASPECTS: DEPENDENCIES

6.1.1 INTRODUCTION

Much of the discussion about methods and effects is in the context of a nuclear plant; however, many of the methods apply to other industries. Dependency is a fact of life, not a specific effect present only in nuclear plants. The impact of dependency between crew members is more noticeable in some industries than others by virtue of their different crew structures. For example, until the aircraft industry decided to modify the relationships between the captain and the second officer, the dependence between the captain and the second officer was very high (i.e., change in the crew factors). In other words, if the captain was wrong, the second officer had little or no chance to modify his thinking or actions.

Need for a human dependency method developed during the early PRA Level 1 studies. The need has become more imperative during the LPSD PRAs, because there are more human actions not backed up by redundant safety systems during these operating conditions. There is a requirement for a method applicable to all aspects of the PRA, including full-power and LPSD conditions.

This section has been written to cover the above. The essence of a dependency analysis is to define those aspects associated with human–human dependencies and build them into a process for predicting the effects of dependence on joint probabilities of related human actions. As can be seen from a study of the other HRA studies (Spurgin, 2000), the following elements are important:

- Cognitive connection between the tasks to be performed
- Time available for actions to be taken
- Relationships between the various members of the crew and support staff in terms of activities to be performed
- Workload (or work stress) during accident
- Connections between tasks as determined by the EOPs
- Spatial relationships between tasks to be carried out from the control board

If the human actions are cognitively connected, then there is likely to be close connection between the events (i.e., strong dependence). Events are cognitively connected when their diagnosis is connected by the same set of plant indications, or when a procedure couples them. Actions stemming from the same diagnosis of plant indications are cognitively connected. Under these conditions, if an error is made in detecting, diagnosing, or deciding, then all of the corresponding actions are likely to be erroneous.

However, if there is a gap in time between the first action to the subsequent actions, then there is an enhanced probability that the mindset caused by the first set of circumstances has changed. Time between actions can be a method for decoupling dependent events. Another method for decoupling events is to use an independent person to review the processes used by the operator and agree or disagree with the action. In the U.S. nuclear industry, the shift technical advisor (STA) was to be used in this manner. Also, other operators may see that the plant is not recovering from the effects of the accident and then start to interact in a positive way with procedure readers suggesting or taking alternative actions. Again, these recoveries are time dependent, and these actions should be built into the dependency model.

The availability of time to enable a change can be manifest in a number of ways, such as the following:

- Original operator has time to reconsider the action taken
- An STA or equivalent person steps in to change the decision
- Other crew members comment on the effectiveness of an action
- Procedures are used to force a review of the action
- Effect is noted of a computer support system continuing to present information on the plant state in that it has not been affected as required by procedure

Most of the dependency models used in PSAs have paid attention to the points outlined above. The approaches used by Swain in NUREG/CR-1278 and the HDT approach are summarized later. Some of the other HRA methods have not explicitly covered dependency, because they grouped all of the aspects of human responses into an upper-level model and thus dependency is accounted for within the model. It has been up to the HRA analysts to base their solutions on the general characteristics of the method they used; for example, using expert judgment when using SLIM (Embrey et al., 1984) concepts.

The analysts' view of operator responses is colored by the model they have in mind. The DT model is introduced here as a way to provide an envelope to contain the domain expert opinions and knowledge to classify and quantify the dependencies between the human actions in the PSA sequences. The elements of the DT system can represent the various influences that affect person-to-person dependency. The DT system is a decision tree structure with the headings corresponding to various dependency influences. It is interesting to note that SPAR-H (Gertman et al., 2004) used a tree approach, but they used Swain's dependency factors as end state probabilities. In other words, they used a method to formalize the selection of the dependency factors, HD (high dependency), MD (middle dependency), LD (low dependency), and ZD (zero dependency) (Swain and Guttman, 1983).

Any accident sequence should be analyzed from the point of view of how dependent human actions are upon each other. The other issue is the incorporation of dependency aspects into the PRA logic structure. In the case of event trees (ETs), this is an integral part of the ET definition. However, in the case of fault trees (FTs), this is more difficult, and care has to be taken to understand the relationships of all human actions to one another. The ETs are structured to follow the accident transient, whereas the FTs present a mixture of top-level events with lower-level events. It just means that the systems analysts and human reliability analysts have to work closely with the plant personnel to understand all aspects of how the crews address accidents in practice. The analysts can be greatly helped in this process by observing how the crews operate during simulated accidents at plant simulator centers.

6.1.2 BACKGROUND

In this section, different approaches to how dependency has been covered in past PRA studies are presented. Following the TMI Unit 2 accident, the USNRC called for each utility to perform a PRA study and present the results to the USNRC. These PRAs were called independent plant evaluations (IPEs), and they needed to be plant-specific studies rather than generic plant studies. Various methods in IPE reports have some things in common. First, the analysts all realize that dependent events can be important in an HRA. There are a number of ways of dealing with dependencies. These are based on Swain concepts or rules, or the analysts can define their own set of rules. The methods used in some U.S. plants are indicated in Table 6.1.

An overview of the methods, with the exception of the Tennessee Valley Authority (TVA), seems to rely on the concept of cognitive dependency modified by time. In other words, the idea that one action is linked to another by the same diagnosis meets the idea that if the first action fails, then the second action will fail. However, there are conditions under which the second action may succeed (e.g., if there is sufficient time between events, if another person is involved, etc.).

Swain also defines his dependencies effect (such as LD, MD, etc.) in terms of the operations carried out by the operators. He identifies the dependency factors in relation to crew, timing, and sequence. This is much the same for the other approaches. In the case of Swain, he gives some general rules. It is then up to the analyst to

TABLE 6.1
Dependency Methods Used for Various U.S. Nuclear Power Plants

Dependency Method	Swain	Analyst Rules	Expert Judgment, Success Likelihood Index Method
Plants	Callaway	Nine Mile Point, Limerick, Robinson Unit 2, and Exelon (ComEd)	TVA, very little information about approach

evaluate the effects and determine the dependency state; see SPAR-H for a way to do this.

In the case of analyst-generated rules, they are similar for cases where the same set of analysts is involved. For example, the set of rules for Nine Mile Point are explicit. The rules for Perry are the same. The HRA analysts for the Limerick IPE and Callaway considered that dependencies are in a state of development. They consider time as affecting the degree of dependence; they also consider the degree of dependency that might exist because of the procedures.

One weakness of Swain's method is the fact that the selection of the dependencies (HD to ZD), are chosen by the analysts. This leads to problems of repeatability and a lack of scrutiny. The analysts, in the case of the Robinson Unit 2 HRA study, have tried to eliminate this aspect by setting up guidelines consisting of 13 rules for the application.

The Exelon (ComEd) Handbook approach is to set up advice for the analyst to consider. The author talks about cognitive dependence, time dependence, cue dependence, and EOP dependence. Unfortunately, the author does not define the effects of these dependent effects and does not give numbers or expressions for the effects.

The DT method builds upon the same background as these other studies. The author tried to take the effects of cognitive connection, time, and independence of operators to build a structure and then suggest what the dependent values might be. The DT approach tries to address some of the perceived shortcomings of the various other attempts to model dependent effects.

6.1.3   CONTROL ROOM CREW PHILOSOPHIES

The effects of crew structure on dependence are discussed below. In the studies, the analysts have used implicit models of crews in deriving both dependence models and the estimation of joint probabilities.

For example, Swain's approach to cover dependent effects within control room crew interactions was based on the old event-based EOPs and knowledge-centered crew responses to accidents rather than the procedure-guided approach currently used. Thus the dependence effects considered in the handbook are based on how crews were expected to respond under those conditions. Some aspects are reasonable, but in general they do not reflect modern control room practice.

Briefly discussed in this section are control room operations in current U.S. plants. In U.S. plants, operations are controlled during accident situations by the use of symptom-based emergency procedures by a licensed senior operator (senior reactor operator [SRO]). Thus the focus of the response is via the activities of the SRO. During quiet conditions, there is at least one operator monitoring the plant condition. In accident conditions, the rest of the crew members assemble quickly.

After the crew has taken the immediate actions called for in the EOPs or symptom-based procedures, they then proceed to the diagnostic part of the procedures. The initial actions are to back up all of the required automatic actions, such as reactor trip. Following the immediate actions, the procedure reader uses the EOPs to diagnose the accident and deal with the identified accident cause. The emergency procedure defines either a specific response or a generalized response, depending on the symptoms associated with the accident. The board or console operators (reactor operators [ROs]) take the series of actions directed by the procedure reader. The crew responses to the accident are monitored by the shift supervisor (SS). A little later in the response, the crew is likely to be joined by a shift technical advisor (STA). The role of the STA is to monitor the plant to ensure that the crew takes the correct series of actions to terminate the accident.

6.1.4 Details of the DT Approach

The approach taken with the DT approach was to build upon prior developments in representing crew actions in responding to simulated accidents (Spurgin and Bareith, 1996). The objective here is to develop a DT type model to be used to predict dependent factors between multiple human actions in accident sequences. A beta factor approach is used here to account for dependencies between a human action and subsequent dependent human actions.

The equation for this is:

$$Pr\,(B|\,A) = \beta Pr\,(B) \tag{6.1}$$

where $Pr\,(B|\,A)$ is the probability of B given that A has occurred, $Pr\,(B)$ is the probability of B independent of A, β is the dependency factor. (β is not the same factor as Swain's, given earlier.)

Figure 6.1 depicts the process of identifying the dependencies between human actions in the cut-sets and estimating the beta factors for the dependent human actions. As can be seen in the figure, the analyst has to determine which human actions are dependent on what other actions in the cut-sets. Once these have been determined, then the process of estimating the dependent effects is considered.

A central part of the process is the construction of a DT structure to be used in the prediction or estimation process, but this is not the only aspect that needs to be considered. The DT construction has to start with an appreciation of the potential causes of dependency between human actions. The discussion in the review section above covers the ground. The analysts constructing the DT need to understand how the crew responds to accidents, because the details of the DT are related to the interactions within the crew following the accident. Even in the case of dependent events, there are potential recoveries or ways in which the crew can mitigate dependency effects.

The arrival of the correct representation of dependent effects is an iterative process. It starts with an evaluation of the contribution of the cut-sets to CDF and the contribution of the various human actions in those cut-sets. For plants other than an NPP, the CDF is replaced by some other loss or damage function, such as the loss of a catalytic ("cat") cracker in the case of a petroleum plant.

The essence of the DT approach is to define those aspects associated with human–human dependence aspects and build them into a process for predicting the effects of dependence on joint probabilities of related human actions. This is similar to the description given in Section 6.1.1.

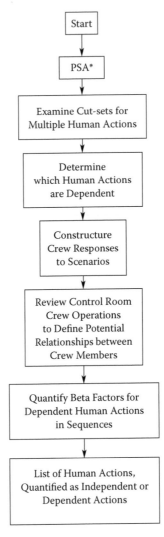

FIGURE 6.1   Process of modifying human actions for dependency (beta) effects.

Figure 6.2 depicts a typical DT for the estimation of the β values related to a set of influence factors. This figure was used in a level 1 PSA for a Vodo-Vodyanoi Energelinchesky (VVER) reactor NPP. It was constructed without input from plant personnel. The following tree should not be considered to be a definitive answer to dependencies for the NPP, because input should have been evaluated by the plant personnel. This version was developed by the author. The balance between the various

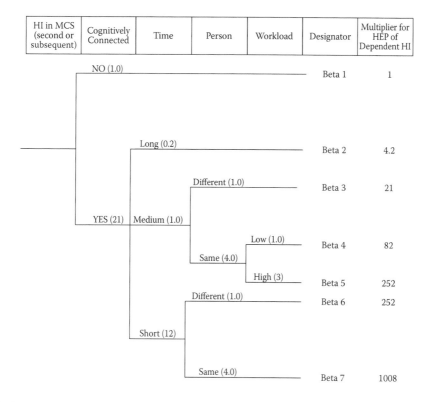

HI in MCS (second or subsequent)	Cognitively Connected	Time	Person	Workload	Designator	Multiplier for HEP of Dependent HI
NO (1.0)					Beta 1	1
YES (21)		Long (0.2)			Beta 2	4.2
	Medium (1.0)		Different (1.0)		Beta 3	21
			Same (4.0)	Low (1.0)	Beta 4	82
				High (3)	Beta 5	252
		Short (12)	Different (1.0)		Beta 6	252
			Same (4.0)		Beta 7	1008

FIGURE 6.2 Typical decision tree for beta estimation.

elements (headings) and their end-state values depends on, for example, training, requirements of plant management, roles of the operators, and procedure design.

The DT structure represents the interactions of the control room crews for the variety of actions taken during the accident sequence and identified within the cut-set. For example, if the immediate cognitive response of the crew to the accident fails, then the set of actions related to that action are also likely to fail. However, the passage of time can ameliorate the situation, and the operators can recover from the situation and take the required actions. The passage of time may not be all that useful if there is the presence of a mechanism to break the operation's mind-set.

If a crew does not correctly diagnose a small break loss of coolant accident (SBLOCA), it is very likely that they will not start the high-pressure safety injection (HPSI) systems. However, subsequently they could depressurize the reactor and later use the low-pressure safety injection (LPSI) system to keep the core covered. Later they could successfully start the heat removal systems to remove decay heat. The DT structure should contain elements representing the various options that can exist over the range of possible accident scenarios. Thus the DT structure should represent the actual plant crews' performance and not some generic model. The dependency model given in Figure 6.2 is an attempt to construct a dependency model for a specific VVER NPP based on the author's understanding of that plant's operations. This model should be reviewed and modified as required based on plant personnel input.

As part of understanding how operators function, it is useful to discuss a simple model of crew performance. One could model human response to accident stimuli in a number of different ways. One way is to build upon a modified stimulus, response, and action model.

Although this is a rather mechanical way of considering humans, it can help one understand human responses to accidents. The version considered here consists of a detection part, a diagnosis part, a decision part, and an action part. This model has been very useful in observing operator responses to simulated accidents (see Chapter 3, Figure 3.4). In practice, it is not possible to easily separate the first three parts to observe each of them separately. Therefore they are easier to handle as one D^3 entity along with an action portion, A. Thus the basis of the dependency model is constructed upon this type of model of human response. It is interesting that SPAR-H uses such a "lumped" model.

Another factor that enters into the process is the ability of humans to recover from mistakes and slips (Norman, 1981). Psychological experiments and observations of human responses tell us that errors of diagnosis are difficult to recover from, even in the presence of contrary information. Often the human mind creates ideas to explain away the counterinformation. This type of error is called a mistake. However, if the operator has the correct intention but executes the wrong action (slip), there is a strong possibility of recovery. This is because the image of what is expected by the human resulting from the erroneous actions does not compare with what actually occurs. Thus the operator can be expected to recover from a slip. In the case of a mistake, there may be no such feedback readily available to the operator to change his mental state. Data taken during a comparison between the use of paper procedures and a computer-based procedure support system confirmed these points (Spurgin et al., 1990b). The STA or safety engineer might be one such mechanism to cause the operator to reevaluate his mental model of what is needed to recover from a mistake.

The D^3 model is a simple model of the cognitive processes of the operator (crew). As such, a mistake can easily be extended to other related actions due to the same set of plant indications or from interpretations of procedural steps. If one considers both the error and the recovery together, errors caused by cognition problems are both important and difficult to recover from. Also, if the operator acts in the short term he may just act on the directions received. However, over time he may reconsider what is happening to the plant and question the decision. Thus time can have an impact, especially if other members of the crew are involved in the process. In the case in which the same person takes multiple actions, there is a strong possibility that if the first action is incorrect, then the subsequent actions are also incorrect. There is a difference between error of intent and slips (correct intent but wrong execution) in terms of the ability of crews to recover from the error.

The different crew structures affect how errors may be recovered. In the case of a strong command structure (i.e., procedure reader driven), errors made by the reader may be difficult to recover from. Three possible recovery paths may occur: cycling through the procedure itself, the presence of an STA or reactor engineer using different procedures, and the board operators commenting on the fact that the plant state is not stabilizing or improving. The DT construct is very much driven by the type of crew organization and the relationship of the crew members to the decision makers.

The process of building the dependency DT has to devolve on plant personnel guiding the activities of the human reliability experts. The plant personnel supply an understanding of how the operators are likely to function during accidents. The human reliability experts can explain how humans are likely to behave. Thus the DT model is a combination of two fields of expertise. In the artificial intelligence field, these are the roles of domain expert and knowledge expert. The first is knowledgeable about the plant-side operations, and the other knows about HRA methods.

The general steps in the process of accounting for dependent effects in the PSA are covered in Figure 6.1. There are two pathways to account for dependent effects. The first is to apply a screening process and then focus on the dominant cut-sets. In this case, the independent and dependent effects are applied at the same time to the human actions in the dominant cut-sets. Once this is done, the results are reviewed to see if any other cut-sets should be added to the dominant set. If there are, then these are recalculated based on actual HEPs and beta factors.

The second way is to apply independent HEPs to all human actions. If this is done, all the cut-sets are reviewed to see if there are any with multiple dependent human actions. For these cases, the analysts calculate the effective beta factors and recalculate the HEPs, and subsequently determine which cut-sets are dominant.

6.1.5 Development of Dependency DT: Process

The process for the development of the DT for dependent human actions is depicted in Figure 6.2. The process consists of using the combination of human reliability (HR) experts and operations experts from the plant to construct the DT. The HR experts should understand how humans behave under plant operational conditions, how to structure the potential operator responses into a logical structure (DT), and how to elucidate information from operations personnel on operator performance. The operations personnel are the domain experts on how the actual crews can be expected to respond in accident situations. These two groups working closely together can produce a reasonable model of crew performance to be integrated into the PSA.

6.1.6 Evaluation of Dependent Events in Sequence

In considering how the process is used, one begins with a review of an accident sequence or cut-set. It does not matter whether the cut-set is dominant or not, but rather that one is going to consider the implications of dependence between the human actions. The accident sequence represents an accident progression. Initiator events lead to accidents, so one is faced with possible operator actions trying to terminate the accident or mitigate its consequences. There is a connection between the crew detecting, diagnosing, and deciding what actions to take, and the actions that they actually do take. This is the dependency in which one is interested. If an action has to be taken within a short time after the crew detects the plant conditions, and the crew fails to detect, diagnose, or act, then the subsequent action will fail. However, if the crew correctly detects, diagnoses, and decides, then the subsequent action is likely to succeed. In the case where the crew correctly understands the accident but fails to take the first action, the subsequent related actions are also likely to fail.

Often, in the cut-set, there are other human actions that may not be connected directly with the first series of actions. These are decoupled from the first series but may have dependent actions within their set of related actions. Under these conditions, the analysts have to consider the use of beta factors applied to human actions with nominal values.

6.1.7 CONSIDERATION OF INFLUENCE FACTORS

Here we consider what factors might affect the dependence of one action upon another action. It is suggested that the HRA analysts and the plant personnel develop their own list along with their potential influence on dependency. Table 6.2 lists the factors considered in the original analysis example.

The first factor in the list is the cognitive connection between the tasks. If the tasks are cognitively connected (i.e., form parts of a complete task related to the same set of plant indications), then the success or failure is linked. That is, until or unless there is a change in the mental state (model) in the mind of the operator.

The time available for actions to be taken during the accident can have a profound effect on the possibility of recovery of erroneous actions. Even in the case of the TMI Unit 2 accident, a recovery action occurred by the intervention of an operator from Unit 1. Unfortunately, the intervention occurred too late to prevent core damage. The mechanisms for recoveries are very important to the evaluation of the beta factors. As can be seen from the IPEs, time can be expected to change the dependency from HD (high dependence) to LD (low dependence) if the time changes from a few minutes to several hours. Thus if only a short amount of time is available, then the possibility of recovery is small; if a large amount of time is available, then recovery is large. Therefore the dependency of one failure on another can be very much affected by the time that is available.

Another feature to be considered is the capability of the crew to recover from errors. As mentioned earlier, the command structure based on the procedure reader controlling the accident response is susceptible to cognitive errors caused by errors in the procedure or the interpretation of procedure statements. Diversity of views of the accident and ways of controlling it can lead to lower cognitive errors.

Figure 6.3 shows that the beta factors are evaluated only for cognitively connected events that fail. These factors do not consider board operations connected

---

TABLE 6.2
Considerations of Influence Factors

Influence Factor	Relative Effectiveness/ Influence on Beta Factor
Cognitive connection between initial and subsequent human actions	Very high
Time available for action	High
Crew member relationships, capabilities affecting dependency	High to medium
Workload of crew during accident conditions	High to low

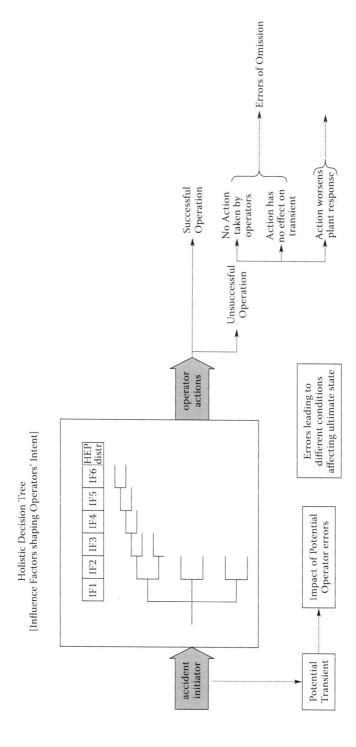

FIGURE 6.3 Relationship between errors of commission, errors of omission, and holistic decision tree model.

to a successful cognitive action (i.e., detection, diagnosis, or decision). It could be enhanced to cover these board-related events.

Time is the next most important effect and is broken down into three bands: long, medium, and short. The branches are arranged in order of increasing dependence, with the top branch the least and the bottom branch having the most effect on the dependency between events. Typical time intervals are as follows: short is from the start to one hour, medium is from one to two hours, and long is over two hours. The actual assessed time might be different for different accident scenarios, but this is for the HRA analysts and plant personnel to assess.

The next effect considered was the effect of other crew members on beta. In the case that time was long, the crew effect was assumed to diminish, and therefore only the medium and short time branches considered the effect of other crew members. Essentially, the presence of other crew members helps the recovery process, and thus the beta value is reduced if there is help from these members. Of course, the help that can be expected is tied to crew structure and training. The capability of the other crew members may be a function of the accident sequence, for in some cases the crews may be very familiar with the transient, and if the main crew member responsible for taking action makes a mistake, they can advise him of the error. However, in other transients, the other crew members may not have sufficient experience to notice and correct errors. The crew structure can affect whether or not the errors are recovered, and if so, how quickly.

The last influence factor is workload. Workload represents the amount of effort that the crew members are involved with during the accident. The workload consists of the plant, reading instruments, using procedures, operating controls, and informing outside agencies. All of these actions can detract from the coordinated recoveries of the crew from errors during an accident. If there is plenty of time, although there might be a lot of activity, it does not continue throughout the transient state; therefore its effect may be ignored. In the case of very little time, the additional workload also does not change the outcome. The effect of workload is only taken into account when there is some time available but not too much.

6.1.8 ESTIMATION OF BETA FACTORS

The next step in the process is to estimate the end states or beta factors for each of the branch paths. In the case of Figure 6.2, there are seven beta values to be evaluated. The first value is set at 1.0 (i.e., no influence of the first event on the subsequent events). The recommended process for estimation of the beta values is to use plant supervisory or training personnel to estimate the ranking of the various end states and then to estimate the actual values. The process for doing so should be a similar estimation of the end states for the DT used in the Level 1 PSA. It is suggested that at least six persons be used and interviewed independently. Subsequently, it might be a good idea to review the final results with all the experts. The results should be discussed with them to check whether there are any last-minute problems with the estimates. An alternative approach is to estimate the branch effectiveness and then use these to calculate the end-state values. In fact, this was the process used to estimate the beta values given in Figure 6.2

6.1.9 COMMENTS

HRA is a developing field, and often the background research and development work has not been carried out. One of the difficult tasks within HRA is the assessment of joint probabilities in the case of dependencies.

Current PSA studies show a general awareness of the dependency effects on human reliability; however beyond that there is no uniformity of identification of the effects and the incorporation of these effects into a process or procedure.

The DT method covered here is a start to try to incorporate identification of dependency influences and assessment of their effects into a logical structure that can be applied by PSA persons. The approach tries to draw upon the experience of plant personnel in a controlled manner and guided by the HRA experts associated with the PRA team.

The DT method discussed here is supported by both plant experience and an understanding of the psychology of control room crew responses. A key aspect of the approach is that it is related to the HDT method used for the estimation of the independent HEP values used in a PSA. It also has face validity. The facts that give rise to dependency are identified along with those factors that lower the dependency effects. The method is a state-of-the-art development in the field and should be an acceptable approach to use to model human dependencies in an HRA.

SPAR-H is an HRA approach used in many USNRC applications and is covered in Chapter 5. One of the details covered there is the SPAR-H approach to dependency, which is partly a Swain and partly a DT approach. The DT part is similar to that given above. DT components are not the same, but the idea of using domain experts to select the pathways is much the same. It appears that many of us have similar ideas and concepts, so maybe our view of the elephant is not fundamentally different.

6.2 ERRORS OF COMMISSION AND OMISSION

There has been a lot of discussion about errors of commission (EOCs) over the last few years, as though there was something magic about them. EOCs are a function of the observer rather than the person taking actions. Except for some cases where persons are deliberately sabotaging a plant, most operators try to take actions that will terminate or mitigate the effects of an accident. Sometimes, because operators make mistakes, the consequences could lead to making an accident worse. Seen from the operators' point of view, they were attempting to correctly respond to the accident.

If one looks at the intent of the operator, his or her aim was to respond to the set of instructions, training, information about the state of the plant seen from the instruments, and the implied instructions from management to correctly act to terminate the accident. If because of all of the influences under which he or she is working, the operator makes a mistake, then the problem is due to these influences.

6.2.1 Development of HDT Model Covering EOCs and OOCs

The end states of the HDT approach can be broken down into four states, rather than just success and failure. The four states are as follows:

1. Success
2. Failure—no action taken
3. Failure—action taken has no effect
4. Failure—action worsens the plant response

See Figure 6.3; the end states (2) and (3) are grouped together as errors of omission (OOCs), and end state (4) covers EOCs.

6.2.2 Errors of Omission (OOCs)

Errors of omission are those actions that appear to have no effect on the plant transient. They include where the operators take no action or where they take an action, but the consequence of the actions is that there is no effective difference between no action and the set of actions taken by the operator. Observations of operators responding to simulated accidents occasionally reveal cases where the operators take actions responding to the accident, but because of incorrect interpretation of the procedures and so forth, they fail to take the correct actions. Also, these actions do not significantly modify the plant response as far as the accident transient is concerned.

6.2.3 Errors of Commission (EOCs)

Errors of commission are those actions that can increase the severity of the accident. Of course, the accident may be made much worse or just a little worse than the case when the operators fail to act.

The mechanism causing the differences rests on the effect of context on the operators. Seen from the outside, failure to control the accident is seen as a failure of the operators, whereas it might be due to a number of reasons, such as poor procedures, inadequate training for this accident scenario, and problems with man–machine interface.

Figure 6.3 depicts the combination of the HDT model along with action pathways leading to success and failure branches. Further, the failure branch is broken into three branches: no action by the operators, no net effect caused by the operators, and worsening of the plant response caused by the operators.

The impact of OOCs, EOCs, and recoveries on the transient response of a plant following an accident is depicted in Figure 6.4. In the case of both OOCs and EOCs, unless there is some additional action by the crew, the plant condition reaches a core damage state. Depending on the actual impact of the EOC, the time to reach core damage may be much less than the situation when the crew does not take actions to terminate or mitigate the effects of the transient (OOC). When the operators take correct actions, the transient occurring after the accident might tend toward the damage state but does not reach it (see Figure 6.4). The actual transient may not be as shown. The curves are illustrative of the concepts.

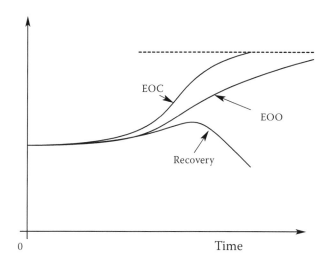

FIGURE 6.4 Typical reactor transient involving errors of commission, errors of omission, and recoveries.

6.2.4 AN APPROACH TO MODELING DIFFERENT END STATES

Observation of operator actions at simulators offers the possibility of gaining enough insight to differentiate between EOCs and OOCs. The problem presented is that on the whole, the operators are very successful and make no errors. This means that the probability of error is quite low; therefore it is not possible to observe many EOCs. OOCs have been seen relatively often. One could argue that as the quality of the various influence factors gets worse, the probability of errors increases, and the proportion of EOCs increases as well. Although this seems to be a reasonable hypothesis, it has to be proven. Reviews of accidents, such as that at TMI Unit 2, indicate that under certain conditions, such as where the training on reactor recovery effects is poor, EOCs can result. ATHEANA (Cooper et al., 1996), along with others, indicated how error producing conditions (EPCs) can lead to EOCs.

Currently, it is suggested that HRA analysts involve the training instructors to estimate for given scenarios the likelihood of EOCs and their relative occurrence. If there is a comprehensive data collection system connected to the plant simulator, then the database could be examined to determine the percentage of EOCs present relative to both OOCs and successful operations.

6.3 EXPERT JUDGMENT METHODS

Expert judgment or expert opinion is at the center of human error estimations. The availability of data specific to a case is difficult to obtain; hence the use of expert judgment is to fill the gap. Even within some HRA approaches that use data sources, like THERP, cognitive reliability and error analysis method (CREAM), NARA, and HEART, there is still a role for expert judgment to select the appropriate task and the corresponding shaping factors and their scales to enable the analyst to fit the circumstances. The approach to using experts is discussed in Chapter 11. One needs to be careful in selecting the experts, to ensure the best estimates of human error probabilities. In Chapter 11, some of the issues and methods used to have some degree of confidence in the process are defined. It is not acceptable for an analyst, who is a knowledge expert but not a domain expert, to guess (estimate) HEPs. He or she needs some help from domain experts. The characteristics of both the knowledge and the domain expert need to be defined. Essentially, the knowledge expert is the HRA analyst in this case, and the domain expert is the person most knowledgeable about the actions of the operators. Operators have often been used for this purpose, but it is my opinion that the better choice is an instructor, because instructors are trying to understand what the operators are doing and why, whereas the operators are taking actions following the procedures and their training. Chapter 11 examines various databases and includes a discussion of expert judgment approaches that have been found useful in carrying out HRA studies. Sometimes HRAs are carried out without a systematic approach to using expert judgments to supply HEP estimates or even weighting factors used to shape HEP estimates. Often it appears that the HRA experts decide on weighting factors without any understanding of how operators respond to accidents.

6.4   COMPUTER-BASED TOOLS FOR ESTIMATION OF HEPs

Computer-based tools have been used for HRA for some time. In fact, there is quite clearly an outcrop of the use of various computer-based workstations used within the PRA/PSA process. Computer capabilities were limited prior to the development of workstations; there were codes for fault trees, event trees, Markoff processes, transient codes for core thermal analysis, fluid system codes for loss of coolant accidents, and so on. There were also codes for the examination of core thermal transients and for particulate dispersion. Early transient codes were developed for control design purposes and for helping with the design of protection systems. Instrumentation systems were designed to detect deviations from normal operating conditions and allowable deviations. Once the deviations were out of the acceptable range, then actions were taken, such as tripping the reactor control rods, operating safety injection systems, and so on. Some of these codes were able to be used to help predict the impact of variations in operator responses. As training simulators have moved from simplified representations of power plants to full scope representations, they have become more worthwhile for training and HRA investigation purposes (see Chapters 12 and 13).

Phenomenological codes for calculation of core, plant, and other characteristics are still used for safety investigations. Many have been improved over time, as computer capacity and speed have improved. In the PRA field, a number of PRA workstations have appeared. Most of the PRA consulting companies developed their own PRA/PSA workstations. Most PRA workstations are based on personal computers (PCs). Some of these are Riskman (ABS Consulting), CAFTA (EPRI), Risk Spectrum (Relcon-ScandPower), and SAPHIRE (USNRC-INEL). There have been a number of changes in the PRA industry in that some consulting organizations have been absorbed into other companies as a result of the 30-year quiescent period in building nuclear power plants. Often, workstations have been developed by one company and supported by another (e.g., CAFTA was developed by SAIC and is now supported by EPRI).

The leading human reliability tool is supported by EPRI. There is a little history behind this development in that an earlier HRA calculator was developed by Accident Prevention Group for Texas Utilities (Moeini et al., 1994) with co-sponsorship from EPRI and Texas Utilities. This calculator was based upon the HCR model, ORE/HCR data, screening values, and an early version of the HDT HRA model for pre-initiator events or latent error HEPs. The objective behind the first EPRI calculator was to record ranges of data for documentation purposes, because the main drive was to base the PRA level 1 on expert judgment using various instructors and other experts. This version of the EPRI calculator was also used by North East Utilities for updates to their PRA.

Subsequently, EPRI funded SCIENTECH to produce the latest version of the EPRI HRA calculator (Julius et al., 2005). The initial version was based on HCR, ORE/HCR, and CBT methods. The current model does not support HCR but covers pre-initiator HEPs based upon ASEP (Sesom. 1987) and THERP. The designers have or are going to incorporate SPAR-H into the code. The objectives of the current designers are somewhat different than those of the original designers in that they are looking to produce a standard approach to performing an HRA study, following the lead of ASME PRA Standard (ASME, 2002). It is not clear at this moment whether HRA has reached a level at which a standard approach is defendable because of issues with data and influence factor prediction.

An independent person, W.D. Lee, placed on the Web a programmed version of HEART (www.tricerote.com/safety/heartcalculator). All of the features of HEART are there, and one can use it to produce HEPs the same as HEART. The only aspect that needs to be improved is to assist the user in the selection of assessed proportion of affect (APOA) values.

The HDT method is basically a computer-based system HRA method. HDT is based on a flexible set of influence factors (IFs) and a range of quality factors (QFs) along with selected anchor values. An Excel program has been written to solve the set of HDT equations. The role of the HRA analyst is to select the balance between the IFs and select the QFs to match the situation determined by the particular accident. The importance of the IFs, QFs, and anchor values can be selected to reflect the actual accident situation. Plant domain experts can be used to select the range of values for every situation. The database is not tied to the method, but the selection of values to evaluate the HEP should be supported by the combination of experience of experts and possible indications from simulators on the sensitivity of particular IFs, QFs, and anchor values. For an application of HDT, see Chapter 10.

It is believed that other organizations have computerized their HRA methods, but it is difficult to access reports or papers on this aspect of HRA.

An adjunct to HRA is the online recording of simulator sessions in which the control room crews respond to accidents. Review of the time data and careful observations of crew actions simulating accidents is very useful to the HRA knowledge expert and can be used to query the domain experts as to what they are basing their judgments on. The knowledge expert can gain deeper insights into crew responses and their possible deviations that can still lead to successful control of accidents. One clear thing that comes out of the simulator results is that not all crews respond in the same time. Also, one crew may respond fairly quickly in one scenario only to be much slower in another case. So a crew may have a "good" response in one case and a "poor" response in another. The time reliability curves can also indicate where crews have problems and where they have none. EPRI funded a number of investigations into data collection systems for simulators. The latest version is CREDIT (Spurgin and Spurgin, 2000). Unfortunately, training staffs generally do not seem to be interested in defining what crews do, beyond brief notes on crew performance. It is interesting that Electricité de France is collecting data (Le Bot et al., 2008) for HRA purposes (based upon the MEMOS method; Pesme et al., 2007). More of this is covered in Chapters 5 and 11.

6.5 FIRES AND LOW-POWER AND SHUT-DOWN PRAs

6.5.1 FIRE PRAs

Fires can occur anywhere within a power plant from different causes. Some fires are due to failure of the staff to remove flammable materials in a timely manner. This is a case of poor housekeeping. Often these materials are used to mop up spilled oil from machinery. Fires can also be caused by electrical shorts due to incorrectly spliced wires and cables. In fact, there are IEEE standards to avoid this type of problem. Electrical fires can also be due to overloads leading to the overheating of cables, which then ignite. Many such potential causes exist. The interest is in how these fires affect the operational staff, and that impacts the safety of the plant?

The two main fire types are those that affect the cables (power and instrumentation) through the plant and those that affect the control room. The first can lead to a reduced capability for the plant personnel to use equipment to shut the plant down (loss of safety equipment) and can present confusing messages to the control room staff if the instrumentation systems are affected. Failure of the instrumentation system can present the operators with no information or faulty information, or even cause the operators to assume a severe accident when in fact there is none.

This is a difficult problem to solve without knowing the physical arrangements of the cabling of a specific plant; an issue in one plant may be a nonissue in another plant. The early regulatory fire requirements were restricted to the loss of function because of the fire. The effects of fires are pervasive and affect the operation of the plant in a number of ways. Older regulations paid particular attention to safety system cabling, and safety groups had to be physically separated. Unfortunately, instrument lines did not need to be separated, and this led to difficulties with the effect of fires on the information available to operators. The identification of cable distribution throughout the plant is a significant task, particularly for the older plants; hence the business of sorting out human reliability effects is equally difficult and is dependent on solving the distribution issue first. The industry (EPRI and USNRC) produced a document in 2005 on the topic of the state of the art as far as fire PRAs are concerned (NUREG/CR-6850 and EPRI, 2005).

The effect of a control room cabinet fire was investigated (Texas Utilities, 1995). The first step in a station response is to evacuate the control room once the smoke becomes dense and chokes the crew, even those using breathing packs. Depending on the design of the control and instrumentation system, it is possible to transfer control to an emergency control room and effect plant shutdown from there. This process has been checked, and as a result, recommendations for additional instrumentation have been made. This is to ensure that the plant is not undergoing another accident at the same time, induced by a control failure due to the cabinet fire.

6.5.2 Low-Power and Shut-Down PRAs

Most of the early PRAs were undertaken assuming that the plant was at or near full power, because this was the typical situation. However, it became clear that the incremental risk of low-power and shut-down (LPSD) could be much higher than at full power. In fact, the risk is higher, but over the years the amount of time that the plant operates at these conditions was significantly reduced, so the overall contribution to risk was reduced.

Why is the incremental risk higher at these conditions? First, the power plants are designed to operate from 100% to 25% load. Therefore all the systems are designed for these situations. This means that the instrumentation accuracy reduces as the flow goes down, for pressure gauges including level instruments operate successfully, unless the ranges reduce to low pressure. Nuclear instrumentation takes care of range changes, but "at power" nuclear instruments have reduced accuracy.

Once the plant drops out of the operating range and moves into shut-down condition, the crews' monitoring capability is reduced. Not only is the effectiveness of the crews reduced, their training in full-power operations is less relevant, the redundancy of the safety systems is reduced, and the relevance of the emergency procedures decreases. For the shut-down case, the main control room is less significant in terms of control. The plant has gone to a different situation; control goes over to a number of persons performing different tasks such as maintenance and testing operations, removing the reactor head, moving fuel in and out of the core, changing pumps, and refurbishing the main turbine. The American Nuclear Society is working on the production of a document to give guidance on the topic of LPSD (ANS, 200x). The USNRC is supporting this work (SECY-01-0067, 2001) and has in the past produced documented evaluations of LPSD operations at U.S. utilities (NUREG-1449, USNRC, 1993).

For these conditions, reliance is on a number of different persons to take actions. Sometimes coordination is via the control room personnel, and other times it is through the plant project manager for these operations. This presents a significantly different task for the HRA specialist, as each state needs to be examined and evaluated. The types of potential accidents are also different and need to be considered. The accidents might include loss of reactor coolant, drop of a heavy load onto fuel elements, and the deboration of reactor fuel pool, so the kinds of accidents are different in the full-power case (100% power). The procedures and the training are also different for these accidents compared with the normal power range of 25%–100% full power. The HRA tools can be used to quantify the risk factors, but even though the experience level of HRA experts is likely to be lower, the domain experts involved in LPSD operations still ought to be able to help in understanding the operations and in giving estimates of human error probabilities.

7 HRA Methods
A Critique

INTRODUCTION

The objective of this chapter is to examine the utility of various HRA methods and models. However, the utility of a particular approach is a function of a number of components, not just the absolute value of a method. The selection of a particular method or model depends on the applicability of the method to the given situation, the ability of the PRA/HRA team to apply the method, and the availability of relevant HRA data. For example, if one is undertaking a completely new application, is there any appropriate data that could be used? In this situation, one's choice might be limited to selecting a method using generic data, even if the source of the data is debatable. Equally, if you are selecting a method to embody many years of operational data, then you are not likely to select a method based upon generic data, because you are trying to capture the essence of the way the plant is being operated.

For cases where data and experience are available, there are a number of approaches that could be used, including the use of expert judgment. When using only expert judgment, one needs to be careful in the organization and selection of the experts. They need to be experts on the topic and not just available engineers. In fact, expert judgment is used in practically all HRA methods in one way or another. HRA methods have changed or been modified over the last few years in response to an awareness of how humans interact with processes. The next section contains a background into the developments in the field and their causes. Many HRA methods were introduced by their developers to respond to identified needs in specific studies. The developers see something they consider to be missing from the then-current methods or something of importance in a particular accident sequence not adequately addressed in the current HRA methods. These aspects then dominate their HRA constructs.

There have been a number of papers and reports covering the range of HRA methods available and criticism of the individual methods. This is to be expected because one has a tendency to fall in love with this method or that one. However, analysts are looking for approaches that are both defensible and usable. More interest in HRA should lead to more people thinking about the pros and cons of the various methods that were developed. A group in England (Lyons et al., 2005) identified some 35 techniques, and there may be even more. This indicates that there has been a reasonable level of interest in the topic over the years.

In order to examine the pros and cons of the various models, a number of experiences can be examined to see whether they could help define the strengths and weaknesses of the various models. For example, if one applies a given method to a variety of plants, both in the United States and in other countries, do the results indicate differences in practice? Also, if one examines the use of a model for a particular case, does the result have face validity, or seem to be accurate? The reviews of the pros and cons are supported by experiences derived from attempting to apply the methods or by observing operator actions at simulators and comparing the results with model indications. For the most part, simulator results do not yield actual failure data, but there can be indications of operator difficulties and their locations.

7.1   BACKGROUND

The background to the development of HRA methods and models is covered in detail in Chapter 5. The objective here is to bring the reader up to speed before launching into a critique of the various HRA approaches. There has been extensive debate among HRA experts about the value of one model versus another. Each HRA model developer has advanced his particular model, and this is to be expected. Without a degree of egotism, the developer probably would not create a model in the first place. In some ways, the model characteristics reflect the needs and experiences of the developer. For example, Swain and Guttman developed THERP (Swain and Guttman, 1983) from the need to model the atomic bomb assembly process and investigate its reliability.

About this same time, a group working on a PRA (Potash et al., 1981) needed to analyze the responses of operators to accidents like steam generator tube ruptures (SGTR) and steam-line breaks. They felt that there was the possibility of confusion on the part of the crew between some accidents, so this aspect was the center of their work.

Later, responding to the perceived shortcomings of THERP, because of its failure to model human cognitive activities correctly, several groups developed time reliability curves (TRCs) based on simulator data. Even Swain developed his own version of the TRC, and this was contained in the update of the draft version of the HRA handbook by Swain and Guttman (1983). Developers of TRC models are listed in Chapter 5.

A variant of the task-related HRA approach, HEART, was developed by Williams (Williams, 1988). This was based on the idea of estimating the probability of a number of standard tasks and having the user try to select the nearest listed task that compared well with the task that he had to quantify. NARA (Kirwan et al., 2005) was developed to replace the HEART tasks with a set of tasks thought to be closer to actual tasks performed by operators at NPPs.

The next change came from a number of different sources. Certain developers felt that there were shortcomings in both the task- and TRC-oriented approaches. They advanced the idea that the context under which the accident took place was more representative of human error generation. Several models have been advanced based upon these ideas. One person in this group is Erik Hollnagel. The others are the developers of ATHEANA (Cooper et al., 1996) and HDT (Spurgin et al., 1999). One must mention, in this context, the work of David Embrey (SLIM), Maud (Embry et al., 1984).

The SLIM approach was based on the idea of using PSFs to identify an HEP within a bounded range. The range was defined by an upper and lower anchor value. Embrey recommended that domain experts be used to select a set of performance factors (PSFs) to predict the HEP. There are some conceptual ideas that are shared by SLIM and HDT. The method of solution is quite different.

7.2   HRA MODELS: CHARACTERIZATION

As mentioned in Chapter 5, a number of different HRA groups have been identified: task related, time related, and context related. Figure 5.1 depicts HRA models that are dominated by one of these characteristics. It is possible that the developers of these models and methods can see something different in their models than the author does.

The models have been grouped into three main groups and have some identifiable attributes that enable one to examine and discuss the basis for the model in a constructive manner. Maybe one could devise a different set of characteristics by which to group the models. However, the idea is to examine these characteristics and see whether they are appropriate to capture the essential features of human reliability. Each of the model types captures some feature that seems important as far as analysts are concerned. In the first group, the idea is that the task being performed is regarded as important and defines the reliability of the operation. The second group is based on the perceived importance of time as the most important parameter in determining the reliability of humans responding to accidents. The last group uses the context associated with the accident to define the failure probability. A short word about context—it consists of the environment under which the crew or personnel operate and it includes the human–system interfaces (HSIs), the knowledge and training of the personnel, and the procedures that should help the crew to respond correctly to an accident. The accident has strong influence on context; for example, it affects how the plant responses are reflected in the HSI.

Chapter 5 covers each of the set of HRA models in some detail, so it will not be repeated here. Of interest is how well each of the methods approximate to actual human performance and help predict the best estimate of human error probability (HEP). Of course, the application of the "best" method may not be possible. The person tackling a particular situation may have to compromise based on availability of data, sufficient time and money, and so forth. Engineering is always about doing the best job under the given circumstances. The opinions of human reliability experts have always been diverse, and in part this is explainable by virtue of the variability of humans. One can test materials to discover their properties, and these properties are repeatable given the same set of circumstances, but people are not materials. They can change their properties by experience, training, and knowledge gained. What they represented yesterday may not be what they are today, so data collected on operators at one plant may not be replicated by data collected at another plant. Investigators are also human and may become fixated on a particular HRA model. For example, we may consider that the task defines the HEP, despite evidence to the contrary. It would be nice if the community of HRA practitioners developed a consensus as to the "best" HRA model to be used at all times, but this may not ever be possible. The best we can do is to bind the error probabilities so that the safety and availability of the plant are not compromised. Often the decision of how to proceed is not reliant upon the HEP value and its interpretation but is much more of an integrated approach with considerations of hardware, economics, and so forth.

7.3 PROS AND CONS OF EACH HRA MODEL

A number of papers have been written covering the topic of the pros and cons of several HRA models; for example, one by Spurgin and Lydell (1999).

Each HRA type has a background of work behind it and has supporters of the approach. The views expressed are those of the author and should be examined critically. As mentioned above, trying to tie down the characteristics of humans is difficult because of the variability of humans. It is likely that there may never be a single comprehensive HRA model that covers all of the analytical requirements demanded of it. Each of the models mentioned has been used in PRA studies, and it is assumed each met the needs of the group applying it. Each developer has considered how to model the activities to the best of his abilities and resources. Perhaps a given method works in one situation, but not in other cases. So to use the words of Swain, there may be no "face validity" in applying a method to cover maintenance operations to respond to the control of an accident by control room staff. The properties of the human reliability "elephant" are difficult to comprehend.

7.3.1 TASK-DEFINED HRA-TYPE MODEL: THERP

THERP is a task-defined HRA-type model. The elements considered are developed by carrying out a task analysis on the specific HRA task to be modeled and quantified. Within the set of subtasks to consider are series tasks, parallel tasks, and recoveries (see Chapter 5). The various subtasks are arranged in a tree form called the HRA event tree (Figure 5.2). Each subtask has a description, and the Swain–Guttman approach is for each description to be compared with a description in the handbook (Swain and Guttman, 1983). Once the nearest description is selected, the handbook gives the corresponding HEP and error factor. This process is then repeated for all of the subtasks. Now, these HEPs are basic HEPs, and in order to make them more representative of the actual situation, they have to be modified by, for example, stress.

So what are the pros and cons of this? On the pro side, it is a straightforward engineering-type approach to a human problem.

If different tasks are being performed, then based upon THERP one would expect that the reliability of humans would mainly be reflected in this difference. The only variable in this case would be the environment under which the tasks were being performed. But this is a lesser effect associated with this method. To test this rule, let us consider a similar plant type around the world. Let us assume that the task is identical for the plants. However, there are some differences in the station's approach to training, selection of personnel, design, and use of procedures. In general, the rules for the human system interfaces are the same or are similar. From comparison between these different plants, one finds the attitude toward the use of procedures is dominant. The operators of the first group of plants are very successful, whereas the operators in the second group are not so successful. The result of this is that the second group's organization changes the whole way the group operates. It would seem from this type of experience that the task is not then the most important element, but just a part of a different set of conditions.

One advantage of the THERP approach is that it fits into the kind of system appreciated by engineers. This means that it has an easy acceptance, which is useful. However, one problem is defining the series of subtasks that contribute to the unreliability of persons executing the task. It is interesting that Swain made the comment that a number of the applications made by others were not right. Perhaps it was due to this particular point. Despite the fact that the handbook is comprehensive, it is still difficult to arrive at a solution that would meet Swain's own review. If the number of subtasks is too high, then the HEP is too high, and if there are insufficient numbers, then the HEP is too low. The skill of the HRA expert is in selecting the key subtasks to model.

Observation of operators in action responding to accidents points to the possibility of preferred error patterns related to problems with procedures and even with incorrect use of board (computer displays) information. This does not seem to reflect the task-defined error model.

Another issue is the supportive data given in the handbook. At the time that the handbook was produced, data could not have come from simulators. It is not clear what the relevance of the handbook data is to even the situation in 1983. Subsequent to the TMI Unit 2 accident, the nuclear power industry changed many things, including placing greater emphasis on training, increased use of simulators, development and use of symptom-based procedures, and upgrades to the main control rooms of NPPs. All of these changes should have had an impact on the errors generated by operators. Therefore one would expect a different error–task relationship.

The advantage of the THERP method is that it can be applied to situations where there are limited or no operating data. Because this seems to have been successful for atomic bomb assembly tasks, the method must be considered appropriate for similar tasks. The method was applied to the movement of spent fuel into a fuel repository for storage purposes. The method was justified for this particular case. Pros for THERP:

1. Appreciated by engineers
2. Relatively easy to apply
3. Can be used for tasks similar to the original application
4. Mostly good documentation

Cons for THERP:

1. Defining the number of key subtasks (really determine human error) may be difficult
2. Data are questionable: Are they relevant to the current situations?
3. Fundamental issue: Is the task the best means to define HEPs?
4. Relating subtask descriptions to current application may be difficult

It is the considered opinion of the author that THERP has a number of significant disadvantages and should not be used except for certain specific applications. It appears that the HRA community has moved away from THERP to other methods. The handbook was a significant contribution to the then-developing field of HRA, but changes have occurred in HRA, and HRA is expected to continue to change. Hopefully, there will be some convergence toward acceptable methods based upon their proven validity.

7.3.2   Task-Defined HRA-Type Model: CBDT

The cause-based decision tree (CBDT) is a development of THERP constructed by Beare and colleagues (1990) acting on a suggestion from the operator reliability experiment (ORE) group to develop a decision tree approach to incorporate results from ORE project. He based the trees mainly on THERP with some modifications based on ORE and his own estimates for HEPs. The subdivision of the trees was based on the experience gained during the ORE work. He devised two main divisions: failure of the plant–operator interface and failure of the procedure–crew interface. Each main group was further divided into four trees. Some discussion of CBDT is given in Chapter 5. As noted, CBDT is one of the components of the EPRI HRA calculator. Pros for CBDT:

1. Most of the pros for THERP apply to CBDT.
2. CBDT is much easier to apply than THERP, in that the tree structure only has to be followed. The user has to consider which branch is most likely.
3. CBDT generally has good documentation.
4. The definition of subtasks is much easier than with THERP—decisions are premade.
5. The HRA event tree is replaced by decision trees (DT), which makes it much easier for the user because one is not tied to generating big trees (see Figure 5.2). Overall, HEP is the sum of DT end states.
6. Codifying CBDT is to assist the HRA user in relating operational experience to the meaning of the headings.

Cons for CBDT:

1. Some users have found the selection of pathways and end states to be limiting.
2. Like a number of HRA methods, the source of the built-in data is questionable.
3. Comprehending the meaning of the headings in the trees is an issue.
4. CBDT tends to isolate the analysts from the real plant experience.

The CBDT approach is an advance upon THERP, and it lends itself to being used successfully, especially as the HRA calculator supports and helps the user. It is difficult to consider CBDT separately from the calculator, because most utility (EPRI) users are going to use the calculator. In the upgrade of CBDT, it is to be hoped that the calculator designers will give some consideration to allowing the use of domain expertise to modify data. Beare realized that THERP was limited in both the subtask definitions and corresponding data due to being part of the ORE team; in fact, he was tasked by the EPRI project manager to try to incorporate ORE simulator data into a form useful for HRA purposes. This turned out to be a difficult job, and in the process of attempting to carry out the task, among other ideas, he investigated Reason's generic error-modeling system (GEMS) approach (Reason, 1987) and held discussions with Electricité de France (EDF) on the database method that they were investigating at the time. Unfortunately, neither investigation worked. He acted upon a suggestion to use the decision tree (DT) approach as a basis to try to incorporate the ORE data into an HRA method.

Beare is very experienced in simulator data collection (Beare et al., 1983), and he previously investigated the utility of simulator data to confirm THERP data during the time he was working with H.E. Guttman. It is not surprising that CBDT was developed drawing upon his experience with THERP and his knowledge of the ORE project. The gaps in the ORE data could be filled with THERP data. The construction of CBDT is a credible achievement that was accomplished in a very short time, and the project cost was quite low. As a side comment, the author took a different path and started the development of the holistic decision tree approach. It would be quite interesting to try to recast CBDT in the light of gaining extensive simulator data from all of the simulators around. Beare had access to ORE data for U.S. NPPs and simulator data from the Taiwan NPP, Kuosheng (Spurgin, Orvis et al., 1990b) and Maanshan (Orvis et al., 1990).

7.3.3 Task-Defined HRA-Type Models: HEART and NARA

7.3.3.1   HEART

Jerry Williams, human factors designer working on the Sizewell NPP in England, proposed a method called HEART (Williams, 1988). His work was based on many years of being involved with human factors. Although the central part of the method is task oriented, the task was defined as something more global rather than the subtask approach taken by Swain and Guttman. His approach was to define a series of generic task types (GTTs) somewhat related to plant operations. These GTTs were defined relatively simply, so the first was labeled "unfamiliar" and the next was labeled "shift without procedure," and so on. The penultimate task was "respond correctly" and the last was a coverall case, "miscellaneous."

For each case, the circumstances were defined, so the first GTT was "Totally unfamiliar, performed at speed with no idea of likely consequences." The next GTT was defined as "Shift or restore system to new or original state on a single attempt without supervision or procedures." The last two GTTs were "Respond correctly to system command even when there is an augmented or automated supervisory system providing accurate interpretation of system status" and "Miscellaneous task for which no description can be found." For all of the tasks, a range of probabilities is given in terms of the 5th percentile, 50th percentile, and the 90th percentile. For example, the range is 0.36, 0.55, and 0.97 for the first task and for the "responds correctly," the range is 6 E-6, 2E-5, and 9E-4.

He introduced the idea of an error producing condition (EPC), and 38 of them were given. These EPCs could be modified by a correction factor to modify its impact. Associated with each EPC was a given value, so some EPCs had a large impact, whereas others had very little impact. The values decreased from the first EPC to the last EPC. So the first EPC was "unfamiliarity." Its description was "Unfamiliarity with a situation which is potentially important but which only occurs infrequently or which is novel." The weighting term was 17. The sixth EPC is "Model Mismatch," defined as a mismatch between the operator's mental model of the accident and that of the designer, and the value given was 8.

The user can modify the impact of the EPC by multiplying by an "assessed proportional of affect (APOA)." The APOA can range from 0.0 to 1.0. Human error probability (HEP) is given in Chapter 5 by Equations 5.1 and 5.2, repeated here:

$$WF_i = [(EPC_n - 1) \times APOA_n - 1.0] \qquad (5.1)$$

$$HEP = GTT_1 \times WF_1 \times WF_2 \times WF_3 \times...etc. \qquad (5.2)$$

where GTT_1 is the central value of the HEP task (distribution) associated with task 1; GTT stands for generic task type; EPCn is the error producing condition for the nth condition; APOAn is the assessed proportion of "affect" for that condition; WFi is the weighting effect for the ith effect; and the selection of the APOA is by expert judgment.

HEART represents a step forward in the task-related methods. Williams tried to generate a more holistic idea of the task. The task is now defined by some of the attributes of the task rather than the operator does this or that. Also, he defines a GTT with a human error probability and its distribution. The other step change is to discuss EPCs and define them. These are conditions that can strongly affect the GTT adversely. He then introduces a term subject to some judgment (APOA) that can lower the impact of the EPCs on the GTT. The user can decide to include as many EPCs as deemed pertinent to the situation, and the EPCs can each be weighted by a different APOA; see Equation 5.2.

Pros for HEART:

1. Appreciated by engineers
2. Relatively easy to apply
3. Experience gained during a number of UK PRAs
4. Reasonable documentation

Cons for HEART:

1. The selection of key tasks is not easy, and descriptions are very vague.
2. HEART identifies some 38 EPCs. Inspection shows that while in the general field of human reliability they may have some meaning, in the field of HRA/PRA they do not apply, for example EPC #38 on age, and there are others.
3. Data for HEPs are of very questionable derivation.

4. APOA is selected by expert judgment, and it is not clear that expertise is available to make these decisions.
5. The fundamental issue is: Is the task the best means to define HEPs?

It is the considered opinion of the author and others that HEART has a number of significant disadvantages; however, at the time of its development, it was good to bring it forward. This appears to be the opinion of some organizations in the United Kingdom, and they have sponsored the upgrade of HEART. The upgraded version is called NARA and is discussed next. One can see the use of expert judgment here. Experts are needed to select the appropriate GTTs, EPCs, and APOAs. This is the same for most HRA methods.

7.3.3.2 NARA

NARA (Kirwan et al., 2005) was sponsored by a UK utility to take the place of HEART and to improve upon some of the characteristics of HEART. The general structure of HEART is accepted, but modifications to the elements in HEART have been undertaken. Some of the shortcomings identified above were identified as reasons to upgrade HEART. The changes to HEART are replacement of tasks by those within NARA, changes in the value of EPCs, and tying the HEP distributions to a new database called CORE-DATA (Gibson et al., 1999). The tasks described in HEART were considered not to be sufficiently close to actual plant tasks, and the HEART tasks were replaced by NARA tasks. The NARA tasks are broken down into four groups: task execution, ensuring correct plant status and availability of plant resources, alarm/indication response, and communications. NARA defines some 18 EPCs, as opposed to 38 in HEART. In addition, NARA introduces a human performance limit value (HPLV), because there is a possibility of repeated human error terms being multiplied together, leading to a very low HEP, and this is considered to be unrealistic. Hence the authors of NARA introduced the process of ensuring that the calculated HEP does not fall below this figure. Those working on NARA were not the only persons to introduce this idea; see Chapter 5.

The general mathematical approach taken in NARA is essentially the same as HEART and is based on the same idea of a series of task reliabilities modified by EPCs, the proportion of which is changed by expert judgment. This is currently in the process of being evaluated by its sponsors, with the possibility of replacing HEART as the HRA method for their PRAs. A valuable step in the process of acceptance is that NARA has been reviewed by an international group of HRA experts (Umbers et al., 2008).

Pros for NARA:

1. Updated version of HEART
2. Relatively easy to apply
3. Based upon CORE-DATA, improvement over HEART
4. Reviewed by an international group of HRA experts
5. Reasonable documentation

6. Can be applied to a situation with limited data; was applied in Yucca Mountain PRA project, selected after careful review

Cons for NARA:

1. Selection of the key tasks is not easy, descriptions are more appropriate than HEART but still not close enough to NPP operations
2. Data that HEPs are based on are better than for HEART, but there are still questions about CORE-DATA data (see Chapter 11)
3. APOA selected by expert judgment, not clear that expertise is available to make these decisions
4. Fundamental issue: Is the task the best means to define HEPs?

Clearly, some of the shortcomings of HEART have been removed, such as the task descriptions and questions about an unqualified database. Also, NARA has been at least peer reviewed. These are all steps in the right direction. There is still a high level of expert judgment.

7.3.4 Time-Defined HRA-Type Model: TRC

The next group of HRA models depends on the concept that time is important and is, in fact, the dominant feature. This seems to be mirrored in one's own experiences. If we have enough time, we can solve any problem. But is this so? A classic case that presents some issues with this condition are exams taken by students. Often in exams the better-prepared students finish the exams before time expires, whereas some of the other students would never finish even given more time. The problem is more a case of preparation of the student rather than time.

In the world of power plants, the case of U.S.-trained crew responding to an anticipated transient without scram (ATWS) is one of high reliability and rapid response. Even other countries using similar training methods and procedure philosophy have equal results. In some other countries, where the importance of the ATWS accident is seen as less important, the results are responded to less reliably and often much slower. This seems to indicate that the amount of training and importance one attaches to the accident is more important than the actual task.

It is interesting to compare ATWS results from two different plants. Both plants are PWRs of Westinghouse design, one is a U.S. four-loop plant and the other is a Taiwan three-loop plant (Orvis et al., 1990). First, the reliability of the crews is pretty much the same as judged by the TRC; there were no failures or very slow responses by either set of crews. There is a range of acceptable times for the crews performing a given task. In fact, for any task there is a natural variation, sometimes short and sometimes longer. One can see these variations from running to intellectual events—the bandwidth of the normalized times is from 0.4 to about 2.5 with the mean normalized time of 1.0.

However, the TRCs were different in that the mean time for the Taiwan plant was larger, but this did not mean that the crew was less reliable. On closer examination of the transient, it was clear that there was a design difference between the U.S. PWR and the Taiwan PWR. The U.S. plant had main control board mounted switches able to shut-off power to the rod drives, whereas in the Taiwan plant, the operators had to leave the control room and go to the auxiliary control room to switch off the power to the control drives. Once the power is removed from the rod drives, the rods drop into the core in about 1.6 seconds, provided that the rods are free to drop.

There seemed to be some confusion about the meaning of the results derived from the simulator results related to operator actions. The time distribution of results represents the time variation of different crews responding to an accident. The response is random. For some accidents, crew A is faster than the mean, and for other accidents crew B is faster. Some crews are faster and some crews are slower, and this can vary from accident to accident. The fact that crews are either slower or faster may not affect whether or not they are more or less reliable than the others. Of course, if the requirement is for the response to be accomplished in a time that intersects within the range of normal crew responses, then the system should be redesigned. This was the point of the original American Nuclear Society Standard N-660 (ANS, 1977). The USNRC funded early simulator-based investigations (Kozinsky et al., 1983) to help establish the criteria for safety-related operator actions (SROAs) to investigate the responses of crews to accidents in support for ANS Standard N-660. The data collected were called nonresponse time data, and the TRCs derived from this data were used in the standard. The data used concentrated upon the successful data, because the interest was whether the crews would be able to deal with an operation successfully. At the time of the study, the state of the industry was that training scenarios were simple, and the emergency operating procedures (EOPs) were event based, so the results are not indicative of what happens now in operations. The work of the ANS was expanded to cover later transients and the use of symptom-based procedures (ANS 58.8, 1994).

The TRCs from the USNRC-sponsored work were used by a number of investigators, such as Hall and colleagues (1982), Dougherty and Fragola (1988), and Hannaman and colleagues (1984b). It was thought at the time that the nonresponse curves were a measure of the cognitive response of operators. The handbook (Swain and Guttman, 1984) first appeared earlier as purely founded on THERP methodology, but due to criticism of a lack of cognitive processing consideration in THERP (IEEE, 1979), Swain introduced a couple of TRCs covering screening and final quantification. These were based upon the TRC concept but modified to contain a tail. It is said that Swain had a session with experts to draw these curves, so in essence the Swain TRCs are based on expert judgment. (The author has been unable to obtain documentation to affirm that this meeting did take place.)

The work of Hannaman and others led to the development of the human cognitive reliability (HCR) curve or correlation (Hannaman et al., 1984b). This is covered in Chapter 5 in some depth. It was generated by the investigators based in part on the simulator studies of Oak Ridge and General Physics (Kozinsky et al., 1983); the work of Rasmussen on skill, rule, and knowledge (Rasmussen, 1979); and the idea that normalization of results could help identify a canonical form (or eigenvalue).

The next step in the TRC progression was that EPRI decided to see if it was possible to confirm or validate the HCR model, so an extensive study was initiated by EPRI to collect simulator data at six collaborating utilities' simulator training centers. The results of this were the operator reliability experiments (OREs) (Spurgin et al., 1990). As far as the HCR model was concerned, ORE proved that not all of the modeling hypotheses were proven. The ORE study provided a considerable amount of data on the responses of crews at a variety of NPPs, both pressurized and boiling water reactors. There were three PWRs and three BWRs. More about the simulator data collections and the processes used is contained within Chapter 12.

The results from the ORE project have led to a better understanding of the responses of crews to accidents and the worth of procedures, training, and good MMI design. The results also indicated that the time response data could indicate that the operators were very reliable or that there was uncertainty and outright failure. The normalized HCR curve for rule-based responses of crews tends to indicate the expected reliable responses of crews to an accident. Deviations from the normalized curve indicate that the crew is having some problems with procedures, MMI, knowledge (training), or even communication protocols. The scenario time is established to enable crews to solve the problems set by the accident scenarios. Sometimes it is clear that a crew has no idea how to deal with the accident or get the plant to a safe shutdown state. Under these conditions, the instructor will terminate the session and discuss the accident with the crew. Sometimes simulator sessions just run out of time and the crew has not succeeded within the allocated time. Simulators are valuable facilities, and time might be limited to half to one and half hours for a given scenario. Crews can fail to complete in time, and this is regarded as a failure. So information gathered can assist the domain expert in the estimation of HEPs. If there are no failures or hesitations, then the crews are very reliable in responding to the accident, so a first estimate might be 1 chance in 1000 (see Chapter 11, Figure 11.1). If there are hesitations, then the crews are less reliable (depending on the number of hesitations, but the crews recover and therefore the estimate might be 1 chance in 100). The last case is the failure rate, the number of failures over the number of crew members, say 1 in 12. The availability of observational data and video recordings can indicate the areas of difficulty with EOPs. A particularly outstanding example was a case where the failure rate was 13 out of 15. Close examination of the records and discussions with the crews (postmortem) indicated an issue with the organization of the procedure statements and a lack of understanding of safety logic. Both issues were corrected by reorganization of the EOPs and the addition of a logic diagram to the EOP documentation. In later simulator sessions (later in the year) for a similar accident, the crews had no failures.

Pros for TRC HRA models:

1. The pros for TRC models relate more to the information gained from simulator-derived TRCs rather than generic TRCs like the curves given in EPRI Report TR 100259 (Parry et al., 1992).
2. Actual TRCs (see Spurgin et al., 1990a) are valuable in the information they yield relative to three things: confirmation of the accuracy and reliability of crews' actions, if there are either hesitancy or failures on the part of crews, and the ability to compare similar tasks but at different plants.
3. They yield insights into why task-based HRA models are questionable.

4. The TRC represents the random variation of crews' responses to accident or any action for that matter.

5. TRCs indicate the importance of training, procedures, and MMIs by the comparison of TRCs for different NPPs. They also show that TRCs can be different due to equipment issues, but the basic reliability of crews with the same training and procedures is similar.

Cons for TRC HRA models:

1. The concept that TRC can be used to define the HEP for tasks in which the available time is large is not supported.
2. The HCR model formulation has significant issues and should not be used.
3. Swain's TRC is not supported by actual TRC results. Operator reliability can be high even for short task times (e.g., ATWS results), whereas Swain's TRC would suggest otherwise.
4. The use of time as the main strong variable defining crew HEP values is not proven. Features such as the quality of the procedures, MMI, and training are more indicative of human error potential.
5. There is a place for TRCs in the set of tools to be used in HRA, but it is more related to simulator data and the insights gained from these sources in carrying out PRA/HRA studies. TRCs can also be of assistance to domain experts in their assessments.

7.3.5 CONTEXT-DEFINED HRA TYPES

As indicated in Chapter 5 and Figure 5.1, there are a number of HRA models that have some closer connection with the context of an accident than either the specifics of the task or time. The list of such models is as follows: CREAM (Hollnagel, 1998), ATHEANA (Cooper et al., 1996), MEMOS (LeBot et al., 1999), SLIM (Embrey et al., 1984), holistic decision tree (Spurgin, 1999), and SPAR-H (Gertman et al., 2004).

7.3.5.1 CREAM Method

The Cognitive Reliability Error Analysis Method (CREAM) (Hollnagel, 1998) was developed by E. Hollnagel based on his experience in the field of both human factors and human reliability. His book covers the field of HRA and includes documentation of his approach to the formulation of an HRA method.

There are two versions of CREAM in Hollnagel's book; one is a simplified view of controlling modes (strategic, etc.), and the other is a more detailed view of human errors. Both methods have been applied in NASA studies. CREAM I was applied as a screening method in an early HRA study for the International Space Station PRA and later was replaced by the HDT approach. CREAM II was applied for one of the HRA studies for the Orbiter PSA/HRA study.

Pros for CREAM:

1. Both parts of the book document versions I and II very well.
2. The database for cognitive failures is based on Beare, Williams, and Swain and Guttman and represents a cross section of available data.
3. He provides a defense for his data in that he says it is there to provide a demonstration of the method, not a justification of its accuracy.
4. He provides for human actions of observation, interpretation, planning, and execution.
5. The relationship between CPC and cognitive processes (CPs) is tabulated, and one can assess the influence on CPs.
6. It provides a worked example of CREAM II in the book.

Cons for CREAM:

1. It is difficult to differentiate between cognitive failures in practice, because the controlling influence now is procedure following, conditioned by training and MMI design and layout.
2. The definition of CPCs should better focus on systematic influences rather than individual influences. It is agreed that the majority of effects are acceptable.
3. Item 5 above may be modified by the characteristics of a particular accident (i.e., influences may change from the ranking given in the table). What is a strong influence in one accident may be weak in another; for example, time may have no influence in one scenario but may be important in another.
4. There is a need to better qualify terms like adequacy of organization. Hollnagel makes a comment to this effect in the text of his book. Also, some of the words used in the table could be better defined. For example, does Institute of Nuclear Power Operations (INPO) give a set of definitions of good to poor organizations in their reports on utility organizations? If this was the case, one could use this as a tool.

7.3.5.2 ATHEANA Method

The ATHEANA method can be broken down into two parts: identification of human errors within an event sequence and quantification of these human errors. The normal HRA methods subsume the errors identified with the event sequence mostly by historical means. The difference is that ATHEANA has a search method to identify error forcing conditions (EFCs) exist that can lead to errors. This type of process is very useful in the case of accident analysis to ensure that all sources of errors are identified. It is helpful to have such EFC taxonomy for future applications.

The problem is how to identify which EFC is more effective than another. One can appreciate that HRA developers consider that EFCs (or EPCs, NARA) can have different weights. One suggestion was to carry out simulator sessions to try to identify the crews' proclivities. This is what has been proposed in the data collection schemes in the past. In the past, this process was carried out implicitly in the PRA/HRA studies by the PRA team members by searching for likely human error pathways. The approach was to use event sequence diagrams (ESDs), and all "major" human events were included. The advantage of the ATHEANA approach is that it should be a more systematic process.

As mentioned before, ATHEANA methodology has been changed from depending on HEART data to using domain experts to generate the HEPs using an elicitation method (Forster et al., 2004).

Pros for the ATHEANA method:

1. Taxonomy for considering EPCs
2. Use of expert judgment elicitation method rather than HEART data
3. Support of the USNRC in the current and future development of the approach

Cons for the ATHEANA method:

1. Need to gain more experience with its application
2. Need to integrate USNRC involvement with Halden Labs and the development of Human Event Repository and Analysis (HERA) database to support, guide, or replace expert elicitation approach
3. ATHEANA could be thought of as a replacement project for the THERP/ handbook but currently does not approach that depth of coverage

7.3.5.3 MERMOS Method

Electricité de France has been involved in undertaking studies in the fields of human factors and human reliability from the early 1980s (Villemeur et al., 1986). Initially, simulator data were collected to understand and improve the human factors design of a workstation for an engineer to support activities during NPP accidents from the main control room. Later, simulator data collected during EDF investigations were used to produce a TRC reflecting this experience. This tool was used during the EDF 900 MWe PWR PSA/HRA study.

EDF designed a later PWR design, called the N4 NPP (CHOOZ NPP). EDF also developed a new operator procedure support system, which is heavily automated. This change led to a reevaluation of the HRA method developed for the PSAs for earlier NPPs. This reevaluation led to the development of the Methode d'Evaluation de la Realisation des Missions Operateur pour la Surete (MERMOS) (LeBot et al., 1999).

MERMOS has been under development for some time. A possible translation of the acronym might be "evaluation method to understand operators' safety mission." The MERMOS method is centered on the use of simulator information and data collected by observers noting the actions of the crew when they are responding to a simulated accident. A number of observers are used, and the data are currently recorded using Lotus Notes. Improvements in the database approach are being considered.

The MERMOS method appears to have a lot of possibilities, because the method makes use of simulator sessions to extract data and insights, which can be used as the basis of inputs into a PSA. The MERMOS user is interested in predicting the probability of crews causing errors. The crews may get into different error forcing situations based on the situation or context. MERMOS merges all of these error probabilities together. The method tries to identify the states by observing crews in operation (responding to accidents at the simulator), by considering past records, and by the considerations of experts. EDF believes that there is a minimum error rate below which one cannot go.

Pros for the MERMOS method:

1. Based upon the use of simulated accidents, helps identify alternative pathways operators might take
2. Uses simulator data to identify performance-related effects
3. Enables analysts to better understand operator action associated with given accidents due to availability of simulator-related information and database of operator actions compiled during various accidents recorded at French NPP simulators
4. Simulators can assist domain experts in estimating HEPs
5. Defines a lower limit for HEPs
6. The MERMOS data can be used to confirm or refute the capability of the EDF operators in being able to deal effectively with accidents. This information can be used to advise station managers and safety authorities of the need for changes to operator training, procedures and instrumentation/displays to enhance plant safety

Cons for the MERMOS method:

1. This response-related model is not founded on a theoretical model of operator responses.
2. Information and database are EDF propriety.
3. Information and database are not transparent to outside persons.
4. Some aspects of the methodology are difficult to understand (i.e., the CICA concept—CICA configurations or orientations of the system, before a situational change).
5. It is difficult to form a clear picture of the method from published papers, and it limits insights gained from MERMOS for outside experts performing HRA studies.

7.3.5.4 SLIM Method

The SLIM method (Embrey et al., 1984) was developed under a contract from the USNRC. A version of SLIM was called FLIM (failure likelihood index method), where the accent was on failure rather than success. It was used by one of the early PRA groups (Pickard, Lowe, and Garrick). The SLIM approach makes use of expert judgment to select a number of PSFs and weigh them according to their perceived contribution in a given accident. These weighted values were then used to derive a modified HEP using anchor values. Another feature of SLIM was the concept of trying to take account of bias in human estimation. In this method, the PSFs are the measures of the influence of the context. So the method relates context to human error. This is not the same meaning of PSF as used in THERP.

Pros for SLIM:

1. SLIM is the first HRA method based on deriving HEPs from context.
2. The method is fairly easy to apply.
3. The method in using NPP expert domain judgments should yield plant-specific HEPs rather than generic HEPs.

Cons for SLIM:

1. Guidance lacking in the choice of PSFs
2. Problems with determining relative importance of PSFs and its basis
3. Issues with PSFs, need to define systematic PSFs rather than all PSFs
4. Difficulty in selecting appropriate domain experts
5. Difficulty associated with selection of appropriate anchor values

SLIM is quite a different approach to HRA quantification than THERP, because SLIM requires the use of domain experts to derive HEP estimates. Apart from the FLIM users, it has not been favored. THERP had the advantage because it was used in WASH 1400, was more appealing to engineers, and was not likely to be subject to as many questions about HEP values by regulators. A test of SLIM-MAUD found that the HEP estimates made by "experts" were inconsistent. One cannot get consistent results using expert judgment methods unless the expert elicitation process is carried out with care. There is always going to be some variation in the estimates, and this should be expected.

7.3.5.5 Holistic Decision Tree Method

The HDT method is described in Chapter 5 and applied in Chapter 10. The method developed out of experience observing control room crew performance during simulated accidents during the ORE project (Spurgin et al., 1990). More details are given in the two chapters mentioned.

Pros of the HDT approach:

1. The method deals with the whole response by the crew to an accident.
2. The method focuses on the context of the accident as it affects the crew.
3. The method is easy to understand.
4. The method indicates clearly which parameters can be changed to improve crew reliability.

Cons of the HDT approach:

1. Expert judgment is needed to ascertain the quality impact of procedures, training, and so forth.
2. It does not explain detailed failures associated with tasks and subtasks.
3. The interrelationships between the various features need to be defined.

4. Anchor values need to be set by expert judgment (i.e., the range of HEPs for the given accident situation need to be fixed, and this sets the possible HEP accuracy).

Behind the context-driven approach is the art of developments in MMI design, EOP layout and test, improvements in training technology, understanding behind communication protocol design, and so forth. Science has advanced over the years in the development of the quality of these fields. The move from one stage of development to another ought to bring increased reliability of human actions dependent on these sciences.

In the last few years, the dynamic test process of usability science has been applied both scientifically and commercially to improve the reliability of getting messages across to users, whether they are housewives using soap products or operators using computer-driven machines. As far as the user is concerned, in these cases, the reliability of the process has improved. These kinds of tests can lead to improvements in MMI for plant control and so on for the rest of the factors that affect the context of the control room crews and others in high-technological industries.

7.3.5.6 SPAR-H Method

The SPAR-H method (Gertman et al., 2004) was developed by Idaho National Laboratories for use by the USNRC; see Chapter 5. The method consists of the combination of a simple human reliability representation based on a cognitive part and an action part with HEPs values associated with each and a set of PSFs to be used in combination with the HEPs. The user determines the human reliability value and uses expert judgment to select the PSFs and their values.

Pros of the SPAR-H approach:

1. Well-documented HRA system
2. Limited human performance choices helps field applications
3. PSFs selections made via tables with guidance
4. Useful examples given in text
5. Method well designed for intended use

Cons of the SPAR-H approach:

1. Simplified human performance approach limits applications
2. Method limits input from domain experts to using tables

7.4 CONCLUSIONS AND RECOMMENDATIONS

A critique has been made of the various HRA models. The level of detail as to the critique could be improved, and it is suggested that this critique forms a basis for a deeper critique by persons selecting an appropriate HRA method to use in their studies. The items in either the "cons" or "pros" lists touch on weaknesses or strengths of the methods. One problem with some methods is the difficulty of selecting a task within the list of tasks thaat compares closely with the actual task being analyzed. One practical way of dealing with not having an estimate of the HEP for the actual task is to examine the range of tasks listed in the selected HRA method and select a couple of tasks that are close to the actual task and then combine the corresponding probability distributions to better estimate the probability distribution to be used the HRA analysis. This brings up the idea of bound databases; these are databases that are part of the model or method. In the case of the CBDT method, the use of the DT approach makes it possible for the end-state HEPs to be modified, and it is believed that some HRA experts have replaced the end states with their own HEPs, more representative of their plant conditions. There is a discussion about the pros and cons of associated data and stand-alone databases in Chapter 11. As some of these methods are applied by different persons, it is expected that they will evolve. One can see the evolution of HEART into NARA. One could expect that NARA, for example, could evolve to encompass more tasks or modify the descriptions to better fit power plants or even to find applications in different fields. The same evolutionary process could be applied to the weighting factors. In the case of the holistic decision tree method, it was found necessary to replace quality factors ranges of good, average, and poor with better definitions based upon human factors considerations. The same was true of QFs for other influences such as training and procedures. The move to do this was based upon trying to assist both the knowledge expert (HRA practitioner) and the domain expert (operator or instructor).

One should remember that it is up to the HRA practitioner to decide if the descriptions of the tasks are good enough. He also has to decide if the use of the combination of task description together with some modification of the basic HEP by PSFs or EPCs is adequate to describe the human activities in responding to accidents.

8 Analysis of Accidents
Various Industries

INTRODUCTION

Much can be learned about accidents that can be applied to a deeper understanding of human responses to accident situations and how they may occur in various industries. Lessons can also be learned from the study of operators' responses to simulated accidents. Some people feel that because simulated accidents are not real, they do not correspond to all of the pressures on the operators and should therefore not be used. But simulated accidents offer a number of benefits, including the fact that one can observe the operators and their actions under a variety of conditions without waiting, possibly for years, for a real accident to occur. Simulated accidents are addressed later in the book.

One of the problems associated with real accident analyses is that often the participants do not live through the experience. Therefore the "view from the cockpit" may be missed. It is much too easy in these situations to blame the operator and put the accident down to "operator error." A big advantage in the use of simulators in studying operator responses is that one has access to the operators and, of course, a large amount of information related to the accident progression. The author's understanding of human responses is colored by participating in many hours of viewing operators' actions during simulator sessions, studying the derived data, and interviewing operators for their insights about what occurred.

In this section, a series of real accidents from different industries are analyzed to show the relationships between the organizations, the design and maintenance of equipment, and humans. One needs to understand the features that go together leading to an accident. Often the environment in which humans operate can cause an accident, or can mitigate the effects of one. The study of these various accidents should lay the foundation for a deeper understanding of these interactions. Probabilistic risk assessment (PRA) or probabilistic safety assessment (PSA) can be used to model accident progressions, but the role of the analyst is crucial to ensure that the elements are mixed in an appropriate manner. The difference between a good and poor cake is the skill of the chef. Automatic processes used to generate risk numbers may have very little credibility relative to reality.

Sidney Dekker has written about the influence of the situation on the response of pilots in his book *Ten Questions about Human Error*, and he says that without understanding the situation from the viewpoint of the pilot, it is difficult to determine who or what is responsible. If one cannot formulate an understanding of the forces at play during accidents, can one really ascribe them to "human error?" Unfortunately, too often the official response to an accident is human error, so then we get into the blame game rather than trying to understand how the accident came about, and then trying to make improvements to reduce the probability of a future accident. An interesting book on aircraft accident analyses was written by Walters and Sumwalt (Walters and Sumwalt, 2000). For each accident, they include the National Transportation Safety Board (NTSB) accident report and the rulings of the Federal Aviation Administration (FAA), their interpretations of the accident, and the responses by the airlines. Anyone interested in the airline business from a safety point of view should read their book.

No person sets out to have an accident. This does not mean that people do not contribute to accidents, but there is usually more to an accident than just the actions of persons.

The purpose of this section is to examine each of the accidents listed to analyze the contributory aspects, so one can see how all these things come together to cause or contribute to the accident in its progression. Contributions can come about by past design decisions, organizational actions and optimization of costs, and by informational limitations and personnel training. In other words, accidents come about due to a number of things.

One thing common to many accidents is the diversity of explanations. In controlled environments, the investigation team can be assembled quickly, and the information disseminated outside of the group is limited; therefore a consistent story can be told. Whether this is correct is another question, but at least there is only one version of the sequence of events. In a number of the cases below, the stories told about what happened vary widely. As a person not directly involved, it is difficult for me to sort out the convoluted stories. A number of derived analyses have been studied and many are questionable, especially because they draw conclusions from clearly unlikely scenarios. Explanations about the sequence of events need to be treated with care. For example, for accidents such as those at Bhopal and Chernobyl, the stories seem to be questionable in certain aspects, despite the effort expended to get to the truth. Kalelkar (1988), in his investigation of the Bhopal accident, said that it is a case of sabotage in that water was injected into the methyl isocyanate (MIC) tank, whereas others have said the accident was caused by the failure of the cooling of that tank. He makes a good case for his conclusion. Yet others have asserted that water was introduced by another mechanism, which on closer examination seems unsupportable. Later, the sequence of events is based upon Kalelkar's explanation, but with a discussion of the relative state of the plant and its design, even though the initiating event was sabotage.

In his paper presented to the Institution of Chemical Engineers conference in May 1988, Kalelkar makes valid comments about the press and the media in the dissemination of the "truth." He points out, quite rightly, that often the more substantiated reports appear later. The public are often unaware of the differences between the first stories and the later deliberations. Until the topic of Bhopal was pursued with some degree of diligence and several "accident" reports were read, no one was aware of the different explanations. Listed in the reference section are several reports relative to Bhopal and other accidents. It might be worthwhile to read them.

A word here about the techniques used to display the accident progression. Event sequence diagrams (ESDs) and event trees (ETs) were used to display the evolution of an accident and its contributory parts. ESDs and ETs are used in the world of PRA and PSA to tie human and equipment failures to the consequences of the accident. PRA and PSAs are discussed at some length in Chapter 3.

The range of accidents covered here includes accidents at nuclear and petrochemical plants, at North Sea drilling rigs, and on railways, and includes both aircraft and aerospace (NASA related) accidents. A few significant accidents in each field have been included for the purpose of showing that accidents are not limited to any particular endeavor. Many accidents could have been used. The ones we have chosen to use here exist in the public domain and memory. One can see that the accidents can occur in "advanced" industries and in industries that have existed for a number of years, like the railways. Table 8.1 contains a list of the accidents analyzed here. Of course, some of these accidents have been discussed many times over; for example, the Challenger accident or the accident at the Bhopal chemical plant. Although some accidents have been discussed in detail elsewhere, here they are dissected, sometimes in ways different than has been done previously. I wanted to consider them as a whole as far as an HRA expert would see them in a predictive manner. The set of accidents here does not cover all types of accidents. Accidents occur on roads between various classes of vehicles; between cars and trains; in hospitals affecting patients (for example, deaths on the operating table, deaths due to incorrect dosage), and in mines. Natural events can also lead to deaths both directly and indirectly; for example, inadequate building practices can lead to buildings collapsing and killing people, specifically during earthquakes.

In the case of a nuclear power plant accident, Three Mile Island (TMI), which was the most serious nuclear accident in the United States to date, is of interest to compare what happened in this accident with what happened in the other significant NPP accident, Chernobyl. While the author was at Westinghouse, consideration was paid during the design of control and protection systems to how one could detect core conditions that could lead to core damage and possible melting. In fact, scenarios similar to that which occurred at TMI Unit 2 were considered. It was decided not to proceed with an in-core temperature detection and display system at the time. It was thought that the operators could detect the condition from the available information and take appropriate actions. What the analysts at Westinghouse failed to realize was that deep knowledge of the transient behavior of the plant was a level of knowledge not available to the operators at TMI Unit 2. However, it should be stated that other plants of similar design to the TMI plant suffered similar accidents but not with the same consequences.

The listed accidents give some perspective on the far-reaching effects of "human reliability," and, of course, some accidents do not necessarily result from direct human actions but are sometimes due to decisions taken by management or during the design or construction process. These accidents will be investigated in more detail later. Often the decisions taken by operators are "forced" by the situation or context under which actions are taken.

The first accidents selected are nuclear accidents, because this is the field in which the consequences may potentially be the greatest. However, the philosophy developed in the United States has ensured that the consequences of an accident can be reduced by prudent consideration of the sequence of events. The early safety philosophy developed

TABLE 8.1
A List of Accidents Covered

Industry	Accident	Consequence
Nuclear	Three Mile Island Unit 2, USA, March 1979	No short-term deaths, at least $2 billion in losses
Nuclear	Chernobyl, Russia, April 1986	Number of short-term (56) and long-term deaths (estimate of 4000); loss of plant and loss of access to housing and land; cost? large
Space	Challenger shuttle accident, USA, January 28, 1986	Loss of shuttle and crew (seven members) shortly after takeoff
Space	Columbia Orbiter accident, USA, February 14, 2004	Loss of Orbiter and crew (seven members) on reentry
Air transport	Tenerife, Canary Islands, Spain, March 1977	Two Boeing 747s collide on the ground, KLM and Pan Am planes, 583 persons killed, loss of both planes, plus ancillary losses
Air transport	JFK airport, New York, USA, 2002	Airbus 300, American Airlines breaks up in the air, November 12, loss of plane, 260 passengers and crew, five others on the ground
Petrochemical	BP Oil Refinery, Texas City, USA, March 23, 2005	14 persons killed, 100 injured, and cost approximately $1.5 billion plus, plant destroyed
Petrochemical	Bhopal, India, December 2, 1984	3000 persons died initially, followed by an estimated 8000 deaths
Oil rig, North Sea	Piper Alpha Gas Explosion North Sea, July 6, 1988	Loss of rig, 165 persons died and 61 survived; cost of the rig and lost production plus fines, additional redesign and training for the industry; insured loss $3.5 billion (estimate)
Railway	Flaujac, France, August 3, 1985	Collision of an express train with a local passenger train, 31 persons died and 91 were injured; the collision took place on a single track
Railway	King's Cross, London, England, November 18, 1987	Fire at a London underground station, 31 persons killed

in the United States was the "defense in depth." This concept was based upon the use of multiple barriers to prevent the escape of radioactivity into the atmosphere. The barriers are the fuel canister, the reactor vessel, and finally, the most important barrier, the containment. The most significant difference between TMI and Chernobyl was the enveloping containment. Later in the post-accident scene, the containment at TMI was bypassed because of an erroneous operation, and it is thought to have led to an unspecified small increase in the death rate due to cancer over time in the affected area near the plant.

The consequences of accidents can range from potential nuclear accidents, which can involve thousands of people (Chernobyl), to aircraft (loss of a plane) covering 2 to 3 people (in small planes) up to 600 people (Airbus 600, for example), and chemical/petrochemical plants involving up to hundreds or thousands due to explosions, fires, or gas leaks (Bhopal).

Petrochemical plant operations show that they have a history of both small and large accidents. For example, the BP refinery at Texas City was reported to have had some 300 accidents prior to the March 2005 accident that killed 15 people and injured many more. The accident is discussed below in more detail and lessons are drawn. These petrochemical accidents do not just occur in the United States, but worldwide. The Flixborough accident in the United Kingdom and the accident in Bhopal in India are notable examples. Some of the accidents are listed in Table 8.2. The Major Hazard Incident Data Service report, HSE/UKAEA (1986) contains almost 10,000 records relating to chemical accidents that have occurred worldwide. The total costs, post-Flixborough, for the top 20 accidents in the United Kingdom are estimated at 430 million pounds (sterling) (1996 values).

Much can be learned from accidents, but often the outcome of post-accident reviews is that the cause of accidents is human error, especially, it seems, when the humans directly involved are killed. Such is often the case in the aircraft industry. The action agent may be the pilot, for example, but why did he take such an action? Can we put ourselves into the situation and determine the pilot's view of the things that led up to his decisions? The accident that occurred in March 1977 in the Canary Islands is a good example of the contributory causes that led to that accident, which involved two Boeing 747s.

In the case of the Canary Island accident, there were a number of contributors: blockage of an access runway by Boeing 747s, language communication difficulties, instruction problems, and even the shape of the runway. All of these played a part, but one could say that the error was caused by the pilot who ran into another 747. Was this pilot error? More to the point, can one predict the probability that the pilot would make this error? The answer is yes if one can understand the environment under which the prediction is to be made. The answer might be that the probability is 0.1 per takeoff. This does not mean that there will be an accident, but in the case of the Canary Island situation, it certainly looked as though the authorities should

TABLE 8.2
Some Chemical Plant Accidents

Plant	Date	Consequence
Flixborough	June 1974	Huge explosion, 28 workers killed
Bhopal, Union Carbide	December 1984	2000+ deaths, settlement several thousands of dollars per victim
Grangemouth, BP	1987, March to June	Explosions, fires, four workers killed, fine of 750,00 pounds
Texas City, BP	March 2005	15 persons killed, 150 injured; BP fined; expected cost $1.5 billion

have reconfigured what was being done at the time, because there was a possibility that the same situation could occur many times due to the number of large planes backed up at the airport at the time. In other words, the probability of an accident is given by 0.1 times the number of repeated situations. There were a number of planes stacked up. There are times when a little forethought in the context of PRA could have obviated the accident. Unfortunately, the management of the situation is often dominated by cost and time pressures, and the consequence of an accident in terms of its cost and delay is not factored into the decision process. The risk–benefit analysis is either incomplete or not even considered. In industrial situations, we see this pattern repeated time and time again in fields as diverse as North Sea drilling rigs (Piper Alpha) to temporary operating fixes at chemical plants (Flixborough).

There have been a large number of rail accidents in various parts of the world. The accidents may lead to people being killed and injured as well as the possible release of toxic materials, which in turn can lead to further deaths and contamination of land and rivers and waterways, depending on the design of the railway network. Since the railway was invented, we have become used to the occasional rail accident and the corresponding loss of life. This is also true of accidents involving other transport-related situations, such as cars and trucks, and, of course, air transport. Our societal response to these accidents can vary, and the media outcry following such an accident can cause an accident to disappear very quickly. They seem to be taken as an immutable fact of life. Railways have always been designed with the amount of traffic in mind, because rail track is costly to build and operate, especially when one considers the need sometimes to select the route to eliminate natural barriers. Often, tracks are designed for a number of different services from fast express trains to local train services to goods trains. The speed differences between these trains can call for care in designing the track and signaling arrangements, most assuredly when the trains can interact. Even in the case of multiple tracks, the possibility of a fast train crashing into a slow train is possible when the trains are on the same track. Things can be worse, such as if there is only one track and the trains are running in opposite directions. In the case of a single track, space for bypassing has to be designed into the track and signaling systems. Like many times in human affairs, we have the balance between cost and safety.

Another issue is that the scheduling of trains can vary during the week. The barriers to prevent accidents on the railway may be broken down into a number of categories:

1. Scheduling the train services in such a way that trains cannot be in the same domain at the same time
2. Providing active controls of the track by signals based upon signal men and backed up by the train driver
3. Prevention of the possibility of trains approaching each other by having a clear space between both trains; this can be accomplished by a passive or active system tied to an awareness of the trains on the tracks

The problem is always about timing, human awareness, and cost. We have a similar situation with aircraft. Optimal loading leads to pressures, plus the organization wishes to keep costs down by reducing labor costs. This can lead to persons working too many hours and becoming fatigued. This, in turn, can lead to errors and accidents.

8.1 THREE MILE ISLAND UNIT 2 ACCIDENT

8.1.1 INTRODUCTION

This accident occurred in Pennsylvania in March 1979 and was the most serious commercial NPP accident to occur in the United States. The accident led to significant damage to the reactor core of the plant. The damage to the core was so significant that the plant has been shut down and is very unlikely ever to be started up again. There was a small release of radioactivity some time after the main accident had taken place. The loss was mainly economic, stemming from the loss of the plant, the cleanup following the accident, and the loss of electricity production.

8.1.2 DESCRIPTION OF THE PLANT

A schematic drawing of the TMI nuclear reactor plant showing the reactor, primary side pump, a once-through steam generator, containment, steam turbine, and the main and auxiliary feed pumps is shown in Figure 8.1. The reactor and plant make up, essentially, a three-loop system. The reactor-side loop consists of the reactor, main reactor pump, and the primary side of the once-through steam generator (OTSG). The reactor heats the circulating water, which is at a high pressure of 2500 psia, maintained by a pressurizer. The heat is transferred to the OTSG, and the cooled water is returned via the reactor pump. In the secondary side, cool water is fed into the OTSG. Here it is heated until superheated steam is formed. The water boils in the OTSG, and the steam so formed then passes to an upper part of the OTSG. The steam continues to be heated until it becomes superheated to about 50 to 60°F. The superheated steam then passes to the steam turbine, which, in turn, drives an alternator, and this produces electricity for the public. The steam passing through

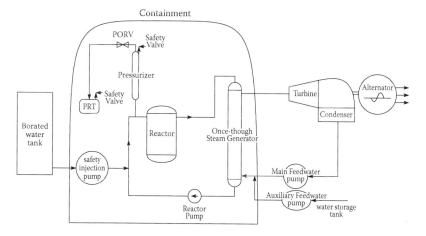

FIGURE 8.1 Schematic of the Three Mile Island nuclear power plant.

the turbine condenses in the condenser and is returned to the main feed pump. The steam in the condenser is cooled by a third loop and is then returned to the secondary loop. This third loop, in the case of TMI, passes the water to be further cooled in the cooling towers.

8.1.3 ACCIDENT: SEQUENCE OF EVENTS

The sequence of events that led to the damage to the reactor core is depicted in an event sequence diagram (ESD) shown in Figure 8.2. The ESD shows the key events that led to the damage to the Unit 2 core, which occurred following a heat up of the primary loop following the loss of secondary-side cooling. The loss of secondary-side cooling was due to the loss of the main feed-water system together with the unavailability of the auxiliary feed-water system. The auxiliary feed-water system was isolated due to a maintenance error. The cause of the main feed does not seem to be very clear. Earlier it was said it was due to the incorrect operations associated with the feed-water filter replacement (i.e., the sequence of actions). Following the loss of feed, both the turbine and reactor tripped. Because the auxiliary feed water was isolated, there was no water replacement for the steam generators. The TMI plants have once-through steam-raising units with very little water capacity, which can lead to a rapid heat-up of the primary side in a situation in which there is a loss of feed water. Due to a normal turbine trip, there would be a heat-up of the primary side together with a rise in pressurizer pressure and level. In turn, the pressurizer pressure-operated valves (PORVs) would lift and steam would be sent to the pressure relief tank (PRT). The design of this tank is such that it should be able to absorb this

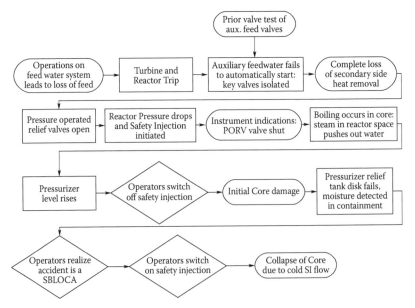

FIGURE 8.2   Sequence of events for the Three Mile Island Unit 2 accident.

energy without lifting a pressure relief disk in the tank. For an anticipated accident of this nature, the plant should be able to ride through the transient.

However, a couple of things were not normal. First, the auxiliary feed-water system was isolated, and second, the PORV did not reseat once the pressure in the pressurizer fell. Eventually, the steam released into the PRT caused the pressure relief disk to rupture, and steam then passed into the containment. Alarms related to high moisture level and radiation should have occurred. These are additional indications of an accident.

The indications available to the operators were taken by them to mean that all was going well; however, the indicators on the main control board only indicated that the signal sent to close the PORVs was zero. This signal did not display the actual position of the valves, so based on this information, the operators did not know that the PORVs were still open. Another piece of information misleading the operators was the fact that the temperature downstream of the PORVs was high, which they attributed either to the prior relief flow or to leakage. Often when a valve lifts, it does not fully reseat, or there may be undercutting of the valve face, leading to a leak.

As a result of the PORVs being open, the primary-side pressure continued to fall until the pressure reached the saturation pressure corresponding to the temperature of the water at the fuel bundle exit. Under these conditions, boiling occurred and steam formed. The steam so formed ascended from the fuel rod assemblies, was captured in the top plenum chamber above the core, and displaced the water, which caused the level in the pressurizer to rise. The pressurizer level continued to rise until it approached the top of the pressurizer. By training, the crew reacted to this state, "to prevent the system going solid," and switched off the safety injection system (SIS).

The heat-up continued and the core was uncovered, leading to clad damage. According to the stories around at the time, a supervisor from Unit 1 arrived in the control room and identified the accident as a small break loss of coolant accident (SBLOCA). Following the identification of the SBLOCA, the actions needed were to reestablish secondary-side cooling and the SIS. These actions were taken; however, the injection of cool water into the core probably caused the final collapse of the stressed fuel bundles and fuel pellets, which fell into a heap at the bottom of the reactor vessel.

The exact sequence is hard to establish, but once the core debris was examined much later, it appeared that core mass exhibited melting, indicating that the effects of decay heating occurred in a closed pile of pellets. Up to now, no radiation was released outside of the containment. The radiation effects of the fuel damage were restricted to the containment. Subsequently, some radiation was released because the containment isolation valves were opened when the containment isolation protection logic associated with the SIS was switched off. Contaminated water was then pumped from the reactor sump to the auxiliary building. Radiation alarms led the operations personnel to close the isolation valves, and the radiation release was minimized.

Like most accidents, the descriptions of some of the details about the accident may differ. For example, reference by Libmann (1996) reflects analyses carried out by the French relative to the accident. This analysis suggests that the formation of void during boiling in the core leads to bubbles being carried around the primary circuit, and this displaces water from the primary system and leads to the pressurizer level increasing. Also, this circulation void finally causes cavitation of the main reactor cooling pumps and subsequently, the operators trip them. The modeling details are different, but the consequences seen by the operators are much the same. Void formation due to boiling in the core leads to a rise of the water level in the pressurizer.

8.1.4   DISCUSSION OF THE ACCIDENT

One can see by examination of the text and the ESD figure that a number of key things did not go as expected. Some were equipment failures, and others were decisions made or not made by the operating crews. The initiating event was the loss of main feed. A number of reasons have been advanced for this loss. One explanation was the incorrect switch-over of main line filters. Water quality is very important for OTSG operation, more so than for the equivalent pressurizer water reactors (PWRs), which have a drum separator system. Other reasons have been advanced for the loss of feed water, but here we are interested in the fact that it was lost.

Once lost, it is difficult to recover because of safety controls that inhibit the fast return of main feed. Given that a reactor/turbine trip occurs following such an action, if the main feed is restarted, it could lead to flooding of the OTSG and the carryover of water into the turbine, which one wishes to avoid. With the loss of main feed, the auxiliary feed-water system should have automatically started, but unfortunately, the isolation valves were left in a closed state after some maintenance testing of the auxiliary feed had been carried out earlier. Therefore no auxiliary feed water was available. It appears that the operators did not see that the isolation valves were left in a closed state.

The reactor system is capable of overriding this condition safely, provided the pressure control system works and water is available from the engineered safety features, including the SIS. However, the PORVs remained open although the indications were that the valves were shut. The control requirement was that the valve should close once the pressure dropped below the set pressure. The control indicator displayed the control output demand (i.e., the valve closed requirement rather than the actual position). This meant that the primary system pressure continued to drop and led to a release of fluids via the PORVs, and eventually to the reactor core being uncovered and only cooled by steam flow, leading to clad damage and the breakup of the core structure.

Often safety and economic considerations overlap, such as is the case here. From a safety point of view one would like to see the feed-water system restarted quickly, because the water capacity of the OTSG is limited to about 15 seconds of full power steaming as compared to a drum-type PWR, which is about 15 minutes. This means that the design of an OTSG dominates the safety aspects for this type of plant. The advantage of the OTSG is the fact that the steam is superheated, leading to improved turbine conditions and performance. There is an economic advantage for these types of steam generator design, but other concerns are brought up.

As mentioned above, the water storage in an OTSG is very limited, so in cases where the feed system does not operate, the crews have to diagnose the problem and fix it quickly. This was not done. The reactor system is capable of overriding this condition provided the pressure control system works and water is available from the engineered safety features, including the SIS. The first stages were carried out and the SIS started; therefore decay heat was correctly removed from the core. However, in this case the PORVs remained open, whereas they should have closed once the transient heat load had been reduced. This meant that the primary system pressure keeps dropping. If the pressure drops too low, eventually the pressure reaches saturation pressure temperature and the reactor water boils. If the PORVs do not close, then the reactor water continues to boil off, leading to the formation of a steam volume in the space above the core. The displaced water then passes to the pressurizer. The result is that the water level in the pressurizer rises. The crew observes the increase in pressurizer level and concludes that there is no longer a need for SIS operation, also according to some analysts, the crews were conditioned by their training to avoid overfilling the pressurizer.

Overfilling the pressurizer means that the pressurizer could go "solid" and the reactor pressure would increase rapidly. The concern was that if the pressure increased rapidly, the reactor vessel could rupture. This understanding was part of naval submarine training, and many operators had received their reactor training in the U.S. Navy. The crew was not aware that the PORVs were still open, and driven by their training, they decided to turn off the SIS. With the loss of both cooling and liquid makeup, boiling continued until the core was uncovered. As the core became uncovered, the fuel elements heated up, leading to damage to the clad. Subsequently, the crew activated the safety injection flow, and this led to the hot fuel elements being rapidly cooled by cold injection water. This is believed to have led to clad failure and the release of the fuel pellets.

8.1.5 ANALYSIS

1. The accident initiator was a result of a poor maintenance operation.
2. The unavailability of the auxiliary feed system was a combination of poor maintenance checks, use of labels that obscured the view of the situation, and the failure of the operators to be aware of the actual state of the auxiliary feed system.
3. Poor design of the PORV control system, where valve status was incorrectly inferred from controller output rather than a direct measurement of valve position.
4. It was reported that a similar accident occurred earlier at one of the Oconee B&W plants and was successfully dealt with. The training department should have made TMI operators aware of this experience.
5. The operators should have been exposed to the transient characteristics of an SBLOCA either by simulator or by class training. Prior to the TMI accident, use of simulator training was very limited, and the capabilities of the simulators were restricted. It is inferred that training of the operators was inadequate to the task.
6. The operators failed to account for high temperature in the PORV discharge line, assumed that temperature was high due to either prior relief action or continuing small leakage post-PORV closing.
7. Key deficiencies included the poor design of the PORV system, inadequate training of staff (operations and maintenance), and inadequate overall control of the plant by management (failure to understand plant, poor training, and lack of involvement in details of the design of the plant).

In Chapter 9, this book estimates the effects of the above deficiencies on a hypothetical prospective analysis. A number of people have seen this accident as mainly an error of commission on the part of the crew. From an outsider's view this is correct; however, the "view from the cockpit" would indicate that the crew was conditioned by its training to carry out the steps that it did. The author does not like the errors of commission and omission characterization. In fact, holding someone accountable for actions taken in line with his or her training is unjustified. If anyone should be held accountable, it is the various organizations coming together to design, build, and operate the plant. The regulators also have some responsibility in the matter.

8.2 CHERNOBYL NUCLEAR POWER PLANT ACCIDENT

8.2.1 INTRODUCTION

The Chernobyl accident took place in Prypiat, Ukraine, on April 28, 1986. The accident led to the complete destruction of the Unit 4 reactor. The reactor was of the Reactor Bolshoi-Moshchnosty Kanalny (RBMK) design. RBMK means in Russian: "large multichannel reactor" (see Figure 8.3). The reactor design is a graphite moderated reactor with cladded fuel cooled by water. In some ways, it is similar to the early British and French magnox reactors, except that the magnox reactors were cooled by carbon dioxide. All of these reactors were physically large. Unlike the U.S.-designed PWRs, none of these RBMK reactors had an enveloping containment but had some reactor building partial containments. The RMBKs had limited containments around the hot leg coolant piping, and as we will see, this was completely inadequate for the actual accident sequence.

8.2.2 DESCRIPTION OF THE PLANT

The RBMKs were designed as both power and plutonium producers. The plant was refueled online, so that the fuel could be optimally used for reactivity control

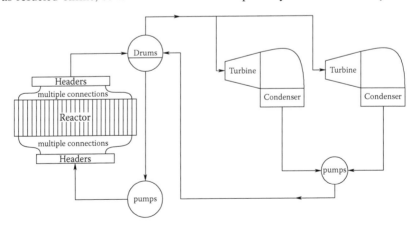

FIGURE 8.3   Schematic of the Chernobyl (RBMK) reactor.

and plutonium production. Part of the design was to use control rods with graphite extensions. This design feature was to help combat the fact that the reactor had a positive void coefficient. In fact, the USNRC does not allow reactors to be built with positive void coefficients. However, the reactivity coefficient for the fuel was negative. Positive reactivity coefficients can lead to runaway power escalations, and this occurred here.

8.2.3 SEQUENCE OF EVENTS

The accident started in a most innocent manner with a test that had been carried out before, but because of delays in arriving at the test conditions, the crew trained to carry out the test was replaced due to a shift change. The test was to see if the delay associated with the diesels coming online could be compensated for by using the main turbine generators to continue to power the reactor circulation pumps. In the previous test, it was unsuccessful, and the voltage regulator design was changed in response to the failure.

However, the experiment was delayed by the load dispatcher because of the need for power. This shift change meant that the persons most knowledgeable about the test requirements were not available. This was the start of the problems. RBMKs are difficult to control, if not unstable under some operating domains, because of the positive void coefficient, which means that if voids are produced, the power increases, leading to more voids (steam) (see Figure 8.4).

Quite clearly from the number of commentaries, the consensus is that the operators were to blame for the accident. Of course, it is easy to blame persons who are no longer able to defend themselves. The author constructed a sequence of events from

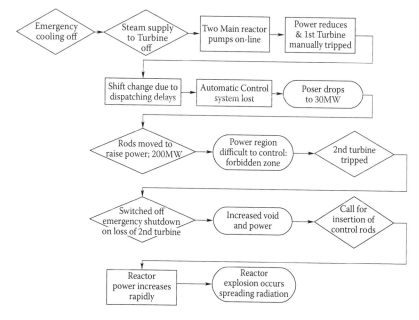

FIGURE 8.4   Chernobyl accident event sequence.

the descriptions available in various documents. In general, the main focus of many documents has been the consequences of the accident (i.e., the number of people who died and the contamination of the soil around the station and beyond, even to the grass in the Welch hills). It is interesting to note that the Russian (Soviet) government said very little at first, and the Swedes were the first to detect the radiation fallout. The operators tried to follow what was requested of them, but they were creatures of the training processes used under the Soviet government.

One could say that the Chernobyl accident was more the result of the design processes and decisions taken in terms of operating the plant than human error, although the actions were taken by the operators. The accident was one that was waiting to happen, if not at Chernobyl, then at some other RBMK plant. It should be noted that even for the other Russian reactors, the VVERs, the latest versions have containments. Also, the training methods being used for the surviving Russian-designed plants have been modified to be in conformance with Western practices. Later in the document, there will be a discussion of the impact of the TMI Unit 2 accident on plant upgrades and training, which could apply to other RBMK, because all of the Chernobyl plants are shut down. Of course, Unit 4 was destroyed.

The test was started with the thermal power being changed to 700 MW from 3200 MW, because the test was to be carried out at a lower power. The crew reduced power too quickly, and there was a mismatch in the Xenon 135 production and decay. Because of the crew's relative lack of knowledge of the relationship between power and xenon poisoning, the crew carried out the operations.

8.2.4 ANALYSIS

1. The experiment was poorly considered from a risk point of view.
2. The replacement crew was not trained in the procedures.
3. There did not appear to be a clear understanding of the dynamics of the core as far as Xenon transients and thermal effects.
4. The crew bypassed safety functions in order to meet the needs of the experiment. This was a clear violation of fundamental plant safety.
5. The plant conditions changed as a direct result of the delay caused by grid distribution requirements. The effect on the plant conditions because of the delay was not factored into the planning of the experiment.
6. Plant management did not appear to be in charge of the plant to ensure the safety of plant personnel and the public. It did not appear that they were involved in the details of the planning and execution of the experiment. If the test was delayed, they should have determined that the replacement staff was not adequate for such a test, despite the fact that an earlier test had been carried out (unsuccessfully).

A hypothetical prospective analysis of the effects on the probability of this accident is carried out in Section 9.4 of the book.

8.3   NASA CHALLENGER ACCIDENT, JANUARY 28, 1986

8.3.1   Background

Space exploration is a risky business for both man and robotic missions—witness the number of failed operations going back to the V2 launches during World War II. NASA's record has been relatively good, although it could be better. Prior to the Challenger accident, there was the Apollo 13 (April 11, 1970) accident involving the rupture of an oxygen bottle, and the oxygen-induced fire during takeoff preparations for Apollo 1 (January 17, 1967). NASA has been a recipient of great expectations with regard to the safety of astronauts. The overall failure rate could be expected to be in the range of 1 in 25 to 1 in 50 launches for the shuttle (Orbiter), despite much care in preparation for the launches. Safety in general is enhanced by the use of redundancy and diversity. Much use of these principles is made in the nuclear business; however, in the rocket field, weight is of great importance, so one has to design highly reliable systems without recourse to multiple backup systems. This means that there is a limit to the reliability of shuttle missions, and any deviation from the basic system design or operation can quickly reduce the safety of the missions. The Challenger and later the Columbia (February 14, 2004) accidents are related to the failure of management to fully understand the impact of design and operational limitations on the basic safety of the shuttle/orbiter.

8.3.2   Description of the Accident

The normal process of launch is for the shuttle to be assembled in a large building (vehicle assembly building) and then transported to the launch site, where the preparations continue, including filling the main fuel tank and ancillary fuel services. The shuttle consists of three main components: the orbiter, the main fuel tank, and the two auxiliary solid fuel boosters (SFBs). The connections between the fuel tank and the three main rocket engines are made earlier during the assembly process. Apart from all of the preparatory processes, which take a large amount of time, the launch process proceeds once a "go" is received from mission control. The main engines, liquid fueled, start, and once thrust reaches a certain level, the SFBs are ignited.

The SFBs are constructed in a number of parts to assist in shipping. There were two versions of the bid for the SFBs: one was the then-current design, and the other was an integral unit. The bidder for the first type won. Perhaps the Challenger accident would not have occurred if the designer of the second design had won. The joint between the components was sealed by a set of two "O" rings and heat-resistant putty.

Behind the earlier launches was a history of bypass burns due to gases from the SFBs affecting the function of the O rings. The design of O ring installation is such that their flexibility is an important feature of their capability to seal the joint. Typically, one would design the installation such that the O ring would be squeezed under normal conditions and once the back pressure occurs, the ring would slide to completely seal the opening. Once the ring loses its flexibility, it then fails to completely seal a joint.

At the time of the Challenger launch, the weather conditions were cold to the point of being icy.

During the prelaunch state, evaluation of the preparedness for launch is made. This is a very formal process, and engineering advice was given not to launch because of the cold weather and the fact that shuttle launch had been delayed for some time under these cold conditions. In fact, there was a criterion based on the temperature of the joint. The engineering advice was based on the experience with the O ring history and the fact that the joint was considered Category 1 as far as safety of the shuttle was concerned. (Category 1 means that failure leads to the loss of a shuttle.) The decision makers were under some pressure to launch because Ronald Reagan was going to announce the "Teacher in Space" program to the nation, implying that the launching of the shuttle was a safe operation. The decision makers thought that it was important for NASA to be seen to be able to launch as and when required. So they approached the manufacturer for a separate assessment, and eventually, despite engineering advice, the representative of the manufacturer said that it was acceptable to launch. It was asserted that there was an implied pressure by NASA on the manufacturer to agree to a launch decision.

The result was that the launch occurred, and a short time into the ascent the O rings failed, and the hot gases from the SFBs impinged on the liquid fuel tank, leading to a large explosion. The Challenger mission was destroyed, with its astronauts, including the teacher, dead. The subsequent investigation stated that the accident was helped by a large wind shear force and the shuttle broke up in part because of the strong wind coupled with the failure of the tank.

Part of the problem was the design of the SFBs. During the firing of each SFB, the whole SFB assembly is affected by the firing of the rocket motor, and this distorts and vibrates. Hence the flexibility of the O rings is completely necessary in the face of the movements of the SFB elements. The rings have to seal effectively and prevent the impingement of hot gases on the liquid fuel tank and some of the support members holding the tank, SFB, and orbiter.

Clearly, this was a case of suboptimal decision making. Of course, the president would have liked to had announced that a teacher had been sent into space, but I am sure that he was horrified that NASA interpreted his priorities and views in this manner.

Later, Richard Feynman (Rogers Commission) stated: "What is the cause of management's fantastic faith in the machinery? It would appear that the estimates of reliability offered by NASA management were wildly unrealistic, differing as much as a thousandfold from the estimates of working engineers" (Feynman and Leighton, 1989, p. 223). The statistical number of accidents to launches to-date of the Shuttles is 2/33 (0.06).

Figure 8.5 shows the various conditions and decisions to the potential for a catastrophic failure of the Challenger shuttle. There appear to be two main features behind the accident: the essential limitations and weaknesses of the O ring design of the joint, and the NASA management's basic belief that the failure probability of the shuttle was very low (see the comment of Feynman). Given the implied pressure stemming from the president's interest in the "Teacher in Space" aspect, coupled with the belief in the robustness of the shuttle, they regarded the caution of the

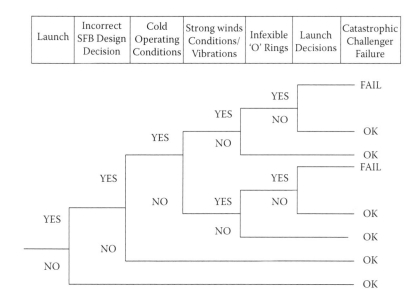

Launch	Incorrect SFB Design Decision	Cold Operating Conditions	Strong winds Conditions/ Vibrations	Infexible 'O' Rings	Launch Decisions	Catastrophic Challenger Failure

FIGURE 8.5 Decision tree associated with the launch of Challenger.

engineers as unwarranted and they decided to launch. However, they still needed to get some outside "expert" to confirm their opinion, and found it with a Thiokol vice president. It has been stated that Thiokol engineers had also informed NASA of the incipient problem and suggested waiting until the shuttle SFBs warmed up.

Figure 8.5 indicates that there are a number of decisions behind the Challenger accident, some associated with the conditions (icy cold weather) and others made much earlier during the design process and cost evaluation stages. One could say that latter decisions dominate. Of course, there are dependences between the various decisions (e.g., cold conditions and O ring flexibility), so if the joint design was not affected by weather changes, then the influence of the cold conditions would be so small as to not be existent. We cannot say that other considerations associated with the overall design would not lead to a loss of a shuttle, as we will see later in the case of the Columbia shuttle accident. In fact, the Columbia accident is similar in a number of ways. It was recognized that there were design weaknesses and some personnel were aware of these in deciding to launch or not to launch. Feynman (1989) in his analysis stated there were two items that lead to incorrect decisions: an incorrect analysis approach that depended on the idea that a successful launch implies that all launches will be successful, and that no degradation of safety controls over time occurs. Both of these have been seen in other circumstances, all too often a situation deteriorates until an accident occurs, and then the management reinstates the correct controls.

The pathway through the decision tree following the "yesses" leads to failure or accident. One of the paths also leads to an accident state, but it is likely to have either a lower probability of failure or the possibility of crew recovery. (I am not sure what recovery methods were available if the shuttle was not too damaged.) Yet other paths lead to OK, but it could be that even in warm conditions, there could be damage to the O rings that could entail some risk to the shuttle. These end states raise the question of potential recovery in partial damage states. Before the shuttle space travel was seen as even more dangerous because of engine explosions, the capsules had escape rockets attached to them in the case of main engine rocket failure. A redesign of the O ring joint seems to be the only clear way to avoid this problem. The current design has increased the number of O rings to three, but it is not clear that this is sufficient.

8.3.3 ANALYSIS

1. Design of the O joint, part of the successful design bid, was not sufficiently robust given its "Category 1" rating and NASA launch needs (i.e., somewhat independent of weather).
2. There seemed to have been an incorrect prioritization of goals. Safety should take precedence over other objectives. It has been asserted that there was a limited safety culture at NASA.
3. The management did not give credence to their engineering staff, which implies a great deal of egotism on the part of NASA management.
4. Management did not seem to have an appreciation of prior issues with O ring blow-bys. The engineering staff was aware of the limitations of the O ring assembly for some time, and nothing was done about a redesign.
5. The engineering staff may not have presented its conclusions on established facts, rather than opinions.
6. Management selected to launch despite the fact that the uncertainty was associated with a Category 1 component or system.

A hypothetical prospective analysis of the effects on the probability of this accident is carried out in a later chapter in the book.

8.4 NASA COLUMBIA ACCIDENT, FEBRUARY 14, 2004

8.4.1 BACKGROUND

Some 18 years after the Challenger accident, another shuttle accident occurred, this time involving the shuttle Columbia. As stated previously, space exploration is dangerous, and it was suggested that the failure rate was 1/25 to 1/50 launches. As a result of the Challenger accident, NASA had made both design and organizational changes in the intervening years. However, the circumstances of this accident look very similar to the Challenger accident as far as decision making was concerned. Maybe change occurred in some parts of the organization but not in others? It seems as though designs were modified, but the attitudes of the decision makers had not changed. It has been asserted, by many persons in the past, that it is easier to change hardware than peoples' opinions. Perhaps all of the press about NASA being a super-organization with state-of-the-art technology was bought into being by the very people who should know better.

Prior to the Columbia launch, there were a number of main fuel tank insulation releases leading to heat tile scratches, and lumps of foam had been seen during launch. These releases were recorded during the ascent phases of the earlier shuttle launches. They should have led to actions by NASA, because they were outside the safety rules for safe operation.

8.4.2 Description of the Accident

Apart from nonsignificant delays, the launch of Columbia was successful. Approximately 82 seconds into the launch a large piece of foam broke off the large external tank and struck the left wing of Columbia, hitting the reinforced carbon heat-resisting panels on the leading edge. Later it was estimated that the foam fragment was about 1 lb plus in weight. It was assessed later to have caused a 6- to 10-inch-diameter hole, which during reentry will allow hot gases to enter into the interior of the wing and destroy the interior wing structure.

The whole of the mission, as far as the astronauts in the Columbia were concerned, went well up to the time of reentry. The NASA management felt that if there was significant damage so as to affect reentry of the shuttle, nothing could be done; so nothing was done despite approaches by engineering staff. The video recording taken during the takeoff operations did not reveal anything wrong; high-resolution film processed later revealed foam debris striking the thermal protection. However, the exact location where the foam struck could not be determined.

The engineering staff asked that steps be taken to carry out a couple of investigations to ascertain the amount of damage by the foam strike. The failure to respond to the engineers' demands was similar to the lack of concern shown by management in the case of the Challenger accident. The engineers requested that various methods be used to assess the damage, including requests to the Department of Defense to use either ground-based or space-based assets to review the Columbia wing. These requests were not acceded to or even blocked by NASA management. Senior NASA managers were of the belief that nothing could be done, whereas the engineers believed that there were a number of things that could be done, from using materials in the Columbia to fill the hole, using the space-walking capability of the Columbia crew, to launching the Atlantis, which could have been readied in the time window to effect a rescue that may have been dangerous. The Atlantis was already on a launch pad and could have been launched without taking risks by hurrying up the preparations.

The mind-set of the top managers appears to have been similar during the earlier accident. They considered that there have been a number of foam strikes before and the shuttles landed without burn-up, therefore all will be acceptable here. If there was a hole in the wing, there is nothing we can do, so why bother to find out definitely if there was a hole in the wing. What a great measure of passivity on the part of NASA managers.

Following the accident, the Columbia Accident Investigation Board (CAIB) was formed to investigate the accident and what actions could have been taken and what actions should now be taken. The CAIB determined that a repair to the wing was possible using materials available in the shuttle. The operation might have been risky to carry out because the astronauts had no training to do this, but they were expected to carry out a similar maneuver closing the fuel doors feeding from the jettisoned fuel lines from the tank, if the door would close. The probability that the fill material would work long enough is a question. The other rescue measure was to launch the Atlantis and transfer the Columbia crew; this is another possibility that would have been tricky, but possible. Because the management did not first check the damage and rule against the engineers' attempts, the whole subject is moot. Under these conditions, no rescue was possible or appears never to have even been contemplated. The NASA management had either written off the Columbia crew or hoped things proceeded well and the Columbia would land safely. One should always prepare for the worst and hope for the best. The words of Winston Churchill might be advice for NASA management: "Never, never, never give up."

As a result of this accident, NASA has taken steps to enable inspection of damaged shuttles in space and to carry out repairs to tiles and broken carbon heat-resisting shields to prevent these kinds of circumstances leading to loss of a crew and shuttle. What other things are still lurking to lead to the failure of shuttles. Feynman was concerned about the main rocket engines, others were concerned about the auxiliary power units, failure of the hydraulic systems, and the possibility of human errors occurring during the reentry and landing phases.

8.4.3 RECOVERIES

The accident sequence is very small for the Columbia accident, if one eliminates the potential recovery actions (see Figure 8.6). The insulation strike occurred, so the only evidence that could be of help would have been if there was no damage. Unfortunately, NASA management hoped that insulation releases would not lead to orbiter damage. Given that there was significant damage, the accident was a foregone conclusion and would end in the deaths of the astronauts and the loss of the shuttle. The failure to collect information on the damage precluded anything being done to help the astronauts. Even if filling the hole had a low probability of success, it was still much greater than the management's position of giving up. The use of the Atlantis, it is believed, would have had a higher probability of success, but who can say without trying. All steps should have been taken, so we could have had the sum of

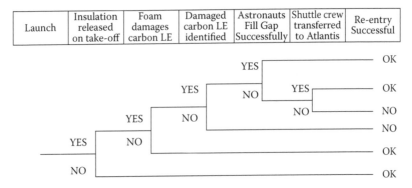

FIGURE 8.6   Decision tree for the Columbia.

successes rather than nothing. The low-cost option was to fill the hole, and if it succeeded, the crew and Columbia would be still here. It is ironic that NASA announced on September 20, 2008 (*LA Times*) that the shuttle Endeavour would be on standby in case the Atlantis astronauts needed to be rescued from space following the launch of Atlantis to service the Hubble space telescope. NASA has now developed materials to repair holes in the carbon heat-resisting shuttle leading edges and replace damaged tiles in space following use of space arms to monitor the surface of orbiters. These types of operation have been carried out recently. In light of early indications of both tile damage and the likelihood of carbon insulation damage, schemes to do what has now been developed should have been acted upon many years ago.

8.4.4 ANALYSIS

1. NASA management failed to aggressively pursue various ideas to rescue the ill-fated astronauts in Columbia. They adopted a fatalistic approach to the situation rather than the "can-do" attitude that prevailed in the case of Apollo 13.
2. Steps have now been taken to shore up the rescue capabilities of NASA.
3. Management did not give credence to their engineering staff in the case of warnings relative to possible damage to the carbon heat-resisting leading edges. They also refused to allow the collection of photos of the suspected damage and to support the engineering staff requests to the U.S. Department of Defense (DOD) for them to gather either satellite or ground-based photos of the damage.
4. Management did not seem to have an appreciation of prior instances with foam shedding and the potential for carbon damage. They seem to be prepared to carry out studies based on nonconservative models on the effect of foam shedding. By not getting the data relative to the actual state, none of their conclusions were based in reality.
5. The engineering staff may not have developed its case for action, but the staff's idea to confirm the situation was correct. If the pictures showed damage, then NASA would have been compelled to take some actions to try to rescue the crew.

8.5 TENERIFE, CANARY ISLANDS RUNWAY ACCIDENT, MARCH 1977

8.5.1 BACKGROUND

This accident affected two Boeing 747s at the Tenerife (Los Rodeos, now Tenerife North airport), Canary Islands, Spain. The planes were diverted because of an explosion at Las Palmas caused by a terrorist group. A number of planes were directed to land at the Los Rodeos airport, and this led to some congestion on the taxiway. The situation improved at Las Palmas, and the planes were allowed to leave. The planes had to be "unpacked" before they could leave. Planes had to reverse-taxi down the runway and then turn on the main runway or turn off a slipway onto the taxiway. The KLM flight was the first

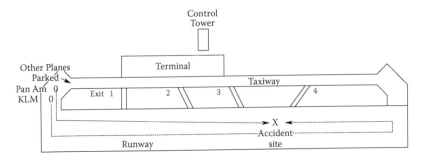

FIGURE 8.7 Depiction of the runway in Tenerife including routes taken by KLM and Pan Am Boeing 747 aircraft.

and back-taxied down the main runway, followed by the Pan Am flight, except the Pan Am flight had to take a slipway/taxiway to the turn-around point. The KLM plane was taking off when it ran into the Pan Am flight taxiing on the same runway but moving in the opposite direction. The majority of the passengers and crews in both planes died. Total fatalities were 583 killed and 61 injured. None on the KLM flight survived.

8.5.2   Accident Location

The layout of the Los Rodeos airport is shown in Figure 8.7. It is a small airport with a single runway and a taxiway. The airport was designed for regional traffic to and from Spain. It was not designed to handle 747s, especially a number of them.

8.5.3   Sequence of Events

Both planes were operating under the control of flight tower personnel. Because a number of flights were redirected to Tenerife, the normal access taxiway was unavailable due to the stack-up. The flight tower directed KLM to proceed via the runway to the far end. Prior to the plane proceeding up the runway, as directed, the KLM chief pilot had the plane filled up with gas to save time later on. Once the plane reached the end of the runway, it turned around and waited for orders on when to takeoff. At about the same time, the Pan Am 747 was ordered to proceed down the runway but was told to take the #3 slipway and then proceed to the end of the runway. At first all was going well. The KLM flight was in position, and Pan Am was moving toward the slipway.

There were communication difficulties between the tower and both planes. The Pan Am crew was told to take the third slipway onto the taxiway. They appeared to have missed the third slipway and proceeded toward the fourth slipway. About the same time, the KLM flight was receiving clearance on what to do once instructions to take off were received. They did not receive the go ahead, but there was some confusion at this point, and they thought they had clearance to take off. The KLM crew did a talkback to confirm the takeoff instruction. In both of these communications, there seemed to be problems between the KLM crew and the tower. Neither party used standard statements, at least according to investigation reports, and this

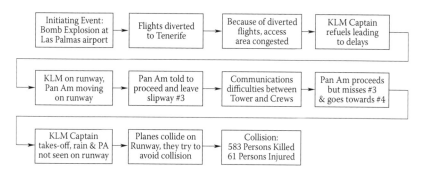

FIGURE 8.8   Event sequence for the Tenerife accident.

eventually led to the accident. In addition, it was reported that there was a "hetero-dyne" beat caused by simultaneous signals from the Pan Am crew. This also added to lack of clarity in the instructions between all parties. This beat signal led to the KLM flight not hearing the control tower statement and also the KLM flight failing to hear Pan Am crew's statement that they were still traveling down the runway. On top of all of this, the weather conditions were poor and none of the crews or tower personnel could see each other. Despite the fact that the flight engineer on the KLM told the captain that the Pan Am may still be on the runway, the captain took off. Just before the collision, both captains saw each other and took evasive action, but it was too late. The Pan Am plane moved off the runway and the KLM plane tried to lift off.

Figure 8.8 shows the events that together led to the fatal accident. Figure 8.9 indi-cates that the accident need not have occurred, and there were several opportunities where if alternative actions were taken, the accident would not have occurred. Two of the events are associated with decisions made by various persons, and the other event deals with communications difficulties. Diverting the flights may not have been pos-sible, but it appears that the organizations involved did not appear to consider the consequences of landing 747s at a regional airport.

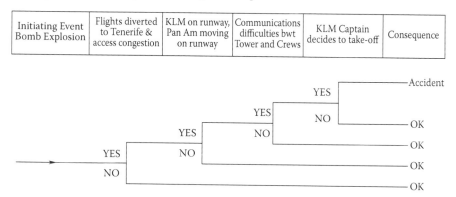

FIGURE 8.9   Sequences of events for the Tenerife accident with alternative paths.

8.5.4 ANALYSIS

1. Following the rescheduling of the planes, the local management should have reviewed the current state of the airport and how the arrival of the 747s could be accommodated and what problems could occur.
2. The impact of the 747 stacking left the availability of the taxiway affected.
3. The management should have carried a risk–benefit analysis to understand what the affect of allowing only one plane at a time on the runway system was compared with allowing movement of two planes.
4. The management should have realized the consequence of language difficulties (American, Dutch, etc., personnel versus usual Spanish speakers) on operations.
5. If one was going to a two-plane moving operation on the runway or taxiway, an additional person or persons should have been positioned on the taxiway to ensure no planes were on the runway at the time that a plane was taking off. This meant that there ought to have been a management control set up to coordinate and control movements.
6. There were no redundant or diverse controls set up in the case of the operations. The management seemed to think that the prior method of operating the airport did not need to change in the face of having a number of 747s flying in. Remember that the airport was a commuter airport.
7. The controlling authority for control of the airspace should have taken steps to assess the situation before allowing 747s to be directed there. This includes suggesting changes in operational procedures to the airport management. It is not clear who the controlling authority was; the probability is that it was the Spanish equivalent of the Federal Aviation Administration (FAA).
8. One can see from the above comments that there was a severe underestimate of organizational difficulties; therefore the prime blame for the accident should have been placed on the management of the facility (local and overall). Planning and organizing the plane movements should have been up to the management. Communications difficulties were just the instruments by which the accident was promulgated.
9. The KLM captain should have been more aware of the actions of others (i.e., movement of Pan Am on the runway) and allowed for additional checks; maybe he was conscious of the time slipping away and this was having an effect on his decision making.
10. Radio communications between 747s appear to have been bad. This raises more concern about the lack of positional information being generated.

8.6 AIRBUS 300 ACCIDENT AT JFK AIRPORT, NOVEMBER 12, 2002

8.6.1 ACCIDENT DESCRIPTION

The American Airlines Airbus 300-600 flight 587 took off to fly to the Dominican Republic shortly after a Japan Airlines Boeing 747 had taken off. Interaction with the wake of the 747 led to the AA 587 first pilot having to take actions to avoid the wake-induced turbulence. In the process of responding to the wake, the Airbus fin and rudder failed. Following the failure, the Airbus crashed, and in the process 251 passengers, 7 crew members, and 5 persons on the ground died. The National Transport Safety Board (NTSB) findings were that the accident was due to pilot error, the American Airlines training program, and structural deficiencies with the fin and rudder assembly of the Airbus 300-600. The NTSB human performance expert presented his view that the first officer responded to the wake-induced turbulence too vigorously, and in part, this was because of the AA training received. The first officer was considered to be a well-qualified pilot. The NTSB formed the conclusion that Japan Airlines 747 was not responsible for the accident. It should be noted that "large" airplanes cause heavy turbulence that can strongly affect smaller planes. The wing-generated vortices can be very strong and persist for a long time. Clearly, this is a major contributory cause that influenced the accident, and how the pilot responded determined the outcome. The closeness aerodynamically of the Boeing 747 to the Airbus 300 influenced the actions of the first pilot without the turbulence. It is likely that the accident would not have occurred. That the first pilot may have overreacted is possible, but the issue of the structural strength of the fin and rudder assembly and the training approach to this type of situation are contributory effects.

The events that combine together to lead to the accident are indicated in Figure 8.10. The initiating event is the aerodynamic proximity of the Japan Airlines Boeing 747 leading to the turbulence affecting the Airbus 300. The pilot's response to the turbulence was in turn influenced by his training, and finally, the pilot's actions, making rapid changes of the control pedals, could lead to larger forces on the fin and rudder assembly than it could take. Once the fin and rudder have broken, the plane would have considerably less stability and crash. Another experienced pilot suggested to me that

Initiating Event 747 Induced turbulence	Pilot's Response to turbulence	AA Training Program	Airbus 300 Fin/Rudder Structural Problem	Fin/Rudder Failure & Crash

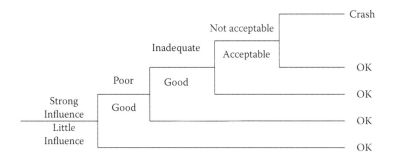

FIGURE 8.10   Event sequence for the JFK accident.

the pilot's training was the primary cause of the crash. The NTSB's investigation and corresponding report along with other reports enable one to understand the accident sequence and corresponding contributory actions and causes. Without this information, the general public would be at a loss to understand accidents, unless one was a member of the investigatory team. Often, the first reports presented by the media are produced without many of the facts being known.

8.6.2 ANALYSIS

1. The turbulence caused by the Boeing 747 was a contributory element in the accident. It does not mean the accident was caused by the 747; however, the closeness of the planes should have been better controlled, so that the vortex strengths are lessened as far as the Airbus 300 was concerned
2. The pilot training seemed to have a strong influence on the Airbus pilot's response to the 747's turbulence and should be modified especially for Airbus 300s.
3. The strength of the Airbus fin and rudder raises some questions about their adequacy, and possibly the fin-fuselage structural strength should be improved.
4. Possibly the pilot training simulator should include an algorithm covering the strengths of various parts of the planes that might be overstressed, so that warnings could be given to avoid certain maneuvers. Like this, pilots would get training on how to carry out operations within the strength envelope of the airplanes.

8.7   BP OIL REFINERY, TEXAS CITY, MARCH 2005

8.7.1   DESCRIPTION OF THE ACCIDENT

The Texas City refinery accident occurred following the restart of an isomerization unit on March 23, 2005, following a temporary outage, when an explosion and fire occurred. The result of the accident was that 15 persons were killed and a large number injured. The accident stemmed from the actions of the startup crew during the filling of a raffinate splitter, which separates oil into light and heavy gasoline. Depending on the reports, the crew did not take action on a liquid level indicator/alarm signal exceeding a set point. This could have occurred because the crew did not observe the alarm or the indicator may have failed. As a result, the crew overfilled the raffinate unit, overpressuring the unit and leading to the operation of a pressure relief valve. Figure 8.11 shows a diagrammatic layout of the column. As can be seen, the level measurement is limited to the lower section of the column. Once the fluid level exceeds the level measurement, the signal does not change, so a review by the crew would indicate the level was still acceptable. However, the crew should have been aware of the fact, as they continued to pump fluid into the vessel and there was no change in level.

Subsequently, seeing fluid was leaving the raffinate vessel via the relief valve, they opened a bypass valve to dump excess fluid to a blow-down drum. The released fluid passed via a heat exchanger, which heated the fluid. In turn, the blow-down tank

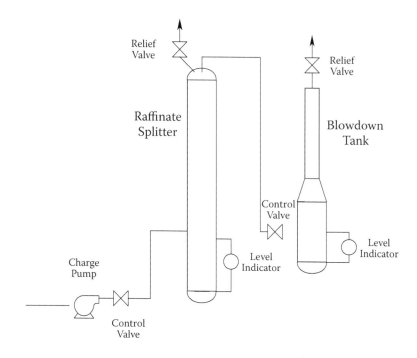

FIGURE 8.11 Simplified schematic of accident at BP Texas City plant.

became overfilled and volatile liquids poured out of the tank and up and over the stack. The combination of liquid and vapor ignited, causing an explosion and fires. The exact cause of the ignition is not known but is thought to be due to a truck with the engine running; however, there have been a number of other ignition sources. The fluid burst into flames and exploded causing the death of 15 persons and injury to a number of others. Many of the people killed in the explosion were in a number of office trailers. If one employed the exclusion zone principle used for nuclear plants, the trailers would not have been positioned where they were, and the death toll would have been much reduced.

Above is a brief outline of the BP Texas City accident. However, it is important to understand something about the accident sequence and its consequence over and above the data provided. The reader is referred to a report by Michael Broadribb (Broadribb, 2006), senior consultant BP, Process Safety, for details and the BP view of lessons learned. For further insights, review reference by a BP audit team following the accident. With regard to the last reference, the team was not very diverse and was made up of members likely to have the same mind-set (i.e., refinery managers from within the BP system).

8.7.2 Accident Sequence

The accident event sequence is depicted in Figure 8.12. The sequence consists of a number of separate phases. The first phase is the overfilling of the raffinate

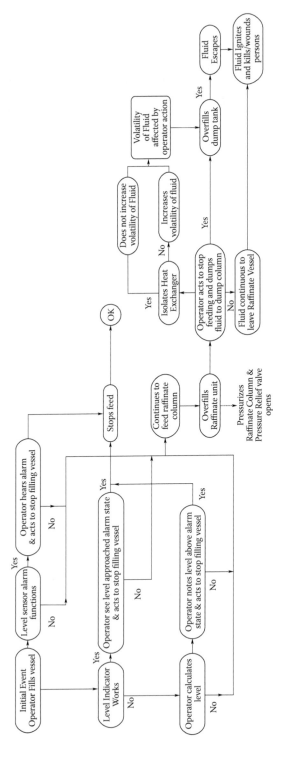

FIGURE 8.12 The BP Texas City accident event sequence diagram.

splitter and fluid release; the second phase is the realization of the release from the splitter column and the opening of the fluid route to the blow-down drum and, subsequently, the release from the drum; and the last phase is the consequence of the release, the explosion, flames, and destruction and death of refinery personnel. Figure 8.12 indicates various pathways that could have led to the termination of the accident and other paths showing the influence of heat exchange processes and the actions by the operators.

In essence, the accident is very simple. The operators' actions led to the release of volatile fluids that caught fire and exploded. There are a number of questions that arise under these conditions that pertain to a number of accidents, why did the accident occur and were the operators solely to blame for this accident? This rationale for causality is very simplistic and on the surface evident, but in practice it is not so simple. Other questions have to be asked and answered, such as what is the reliability of the plant and its instrument systems; is the information presented to the operators both clear and correct; are the operators taught to take the appropriate actions given the situation; does the design of the plant cover the range of possible accident conditions, for example, is the blow-down drum capacity sufficient to contain the whole contents of the raffinate column? The author has drawn on the lessons learned from the accident by BP personnel and several are listed in the next section.

8.7.3 ANALYSIS

1. There were deficiencies in the plant maintenance and the quality of the plant inspections was not very good. Key roles of humans in both active control and maintenance are underestimated
2. The control equipment was unreliable and inadequate. For example, if the level alarm was achieved by a single detector and there was no redundancy, this is not good. In addition, there was a need for more level measurements in the raffinate column.
3. The isolation valves were either nonfunctional or inaccessible.
4. Key buildings were used by controllers and other personnel and were not well located and were lost during the accident.
5. One thing occurring regularly seems to occur in the chemical business— the risk of plant modifications is underestimated. The interactive effects between plant systems are underevaluated.
6. There are gaps in coordination between local and top management with respect to the operational safety of the plant. There is a need for the organization to carry out far more comprehensive safety analyses involving both plant operations and equipment design.
7. The application of PRA and HRA techniques should be carried out to enhance the safety of the plant and its operation. Risk-based methods are being used to enhance the safe and economical operations of nuclear plants, and those could be usefully applied to petrochemical plants.

8.8   PIPER ALPHA OIL RIG ACCIDENT: NORTH SEA, JULY 6, 1988

8.8.1   Introduction

The North Sea gas fields have provided the United Kingdom with the means to supply energy to replace the dangerous coal fields in the generation of electricity and
also supply gas for heating purposes. The fields were being exploited by companies
with knowledge of developing offshore gas and oil fields based upon rigs. American
experience in the Gulf of Mexico provided the engineering knowledge to build these
rigs. One such rig was the Piper Alpha platform operated by Occidental Petroleum,
and it was a significant node in the gas collection system. This is of significance
because the fault leading to the accident was compounded and enhanced by these
other rigs feeding into Piper Alpha. The initial fire led to explosions caused by the
failure to quickly isolate these other sources. Piper Alpha accounted for around 10%
of the oil and gas production from the North Sea at the time of the accident. The
platform began production in 1976 first as an oil platform and then later converted to
gas production. An explosion and resulting fire destroyed it on July 6, 1988, killing
167 men. Total insured loss was $3.4 billion. To date it is the world's worst offshore
oil/gas disaster. The accident was investigated by the Lord Cullen Enquiry and the
report was submitted to the UK government in December 1990.

8.8.2   Description of the Accident

This accident, like many accidents, started quite innocently. On July 6, 1988, work
began on one of two condensate-injection pumps, designated A and B, which were used
to compress gas on the platform prior to transport of the gas to the Flotta site on land.
A pressure safety valve (PSV) was removed from compressor A for recalibration and
recertification, and two blind flanges were fitted onto the open pipework. The dayshift
crew then finished for the day, before the pump was returned to service. Documentation
for the pump and the valve were separated and kept in different places.

During the evening of July 6, pump B tripped, and it was not possible to restart
it. The nightshift crew decided that pump A could be brought back into service
after review of paperwork. It appears that the crew was not aware that the PSV was
removed and replaced by blank plates. Once the pump was operational and the pressure in the line increased, gas leaked from the two blind flanges, and, at around 2200
hours, the gas ignited and exploded, causing fires and damage to other areas with the
further release of gas and oil. Some 20 minutes later, the Tartan gas riser failed, and
a second major explosion occurred followed by widespread fire. Fifty minutes later,
at around 2250 hours, the MCP-01 gas riser failed, resulting in a third major explosion. Further explosions then ensued, followed by the eventual structural collapse of
a significant proportion of the installation.

8.8.3   Description of Accident Sequence

Figure 8.13 shows the sequence events leading to the loss of life and the destruction
of the rig. Some of the decisions depicted on the figure are ones taken much earlier

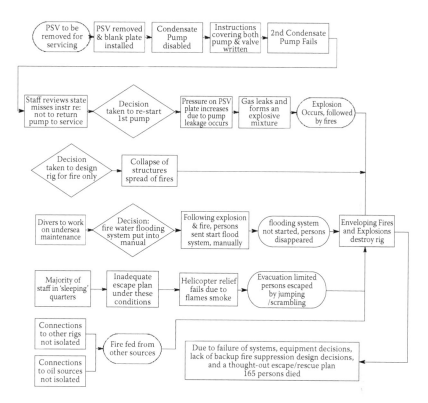

FIGURE 8.13   Event sequence diagram for the Piper Alpha disaster.

by management and rig designers. The rescue of personnel after the initial fire has taken place is affected by the design of the rig and of the rescue vessels. The whole accident reveals many shortcomings, from personnel training, procedures, and operational rules to escape plans, rig design, and even the design of escape ladder deployment on the rescue boat. The Cullen report was written in two volumes, one covering the accident sequences and issues and the second concerned with recommendations for change to enhance the overall safety of rigs. Interestingly, he pointed toward the nuclear industry and the use of PRA and HRA techniques used in design. Actually, he referred to the use of fault trees and human factors, but his comments should be better interpreted as risk-based evaluations of both design and operation.

The top part of the sequence diagram shown in Figure 8.13 covers the basic accident from removal of a PSV through the failure of the second pump (B) and the decision to restart the first pump (A). This operation then leads to the initial fire and explosion. The next branch shows the effect of designing the rig for oil operations and failing to reconsider the design of the rig if there is a change to gas production. For oil production, the fire wall designs were acceptable because they were designed for oil fires. For fire walls for gas production, one needs to consider both fires and explosions. The collapse of the walls led to the spread of the fires and the undermining of steel structures and pipes. This meant that gas being supplied from the other rigs played a major role in the further destruction of the rig and affected the safety routes of the rig personnel.

The next branch of the sequence diagram deals with the cause of fire suppression water. The rig was designed with water sprinklers in place to deal with fires; however, because there were divers working on the platform substructure, the sea water pumps were placed into manual operation as a precaution to safeguard the divers in case the pumps were spuriously turned on. During the fire, persons were dispatched to turn on the pumps manually but were not able to gain access to the diesels to start them, and the water flooding of the rig was not achieved.

The next branch deals with why the death toll was so high. The majority of the crew was in sleeping quarters and the escape plans were inadequate. Helicopter relief failed because of smoke and flames. Some persons were forced to jump into the sea in their attempts to escape and a rescue boat even sank. Even the large firefighting and rescue ship had a design flaw that prevented the escape gangways being extended quickly enough to save lives.

The last part of the figure deals with the gas connections to other rigs. These were still supplying gas to the stricken Alpha rig long after the start of the problem. There was a failure of management to take action quickly to shut off the pipes in time and even to be aware of the issues associated with the storage of the gas in the pipes. Isolation valves needed to be close to each of the rigs. Oil from one of the rigs was stopped after another explosion, and the large rescue ship was driven off by the heat given off, and the heat led to the melting of the structure.

The final result was the loss of 165 persons and the complete destruction of the Piper Alpha rig. Many of the shortcomings of the design and operation were known. Even action by other rig supervisors was not taken early enough because of the cost of restarting facilities. The total insurance cost of the accident was estimated to be $3.4 billion in 1988 dollars. The recommendations by the Cullen Enquiry contained in Volume II of the report were accepted by the industry. Volume I of the report was the detailed review of the accident. The above description cannot really replace the work of the team working on the analysis of the accident as reported in Volume I of the Cullen Report.

8.8.4 Analysis

1. The report shows that there can be close connections between operations and maintenance. Communications and indications as to the state of equipment have to be clear. The later shift operators were not aware that the PSV had been removed. Information about its state was contained in a different location than that referring to the pump.
2. There was a failure to upgrade the fire protection walls as a result of moving from oil to gas operations. This shows that there was a significant lack of a safety culture in the management of the rig.
3. It was quite right for the divers to have protection against spurious operation of the fire water flooding pumps; however, remote operation of the diesels should have been designed.

4. The loss of the control room due to the fire or explosion should have been foreseen, and an auxiliary control center ought to have been set up in a safe environment.

5. Procedures for the operation of fuel shut off should have been developed with the support of management, and the conditions for their operation should have been clearly defined, and it should have erred toward the safety of the personnel. Often safe operations also lead to better operations and the lowering of economic risk.

6. Evacuation paths and sheltering of the crew did not seem to be satisfactory, because many of the personnel could not evacuate except by dropping some 200 feet into the sea. This is not a very good solution; it might be a case of choosing how one wishes to die. There ought to be multiple escape routes so that the crew could leave safely wherever the fire location.

7. Sadly, despite the arrival of a ship designed to suppress fires and rescue personnel, it was not capable of functioning to perform the latter task. Surely it should have been tested under realistic situations before being commissioned for firefighting.

8. Lord Cullen was right when he said that PRA/HRA techniques should be used, but the real problem was that top management had very little will to use such techniques, because they appeared to be unaware of the risk–benefit from safe operations and the need for them to develop a safety culture. The question here is, can they afford to incur a loss of $3.4 billion and the loss of 167 skilled men by not paying attention to maintenance, good design, and careful and safe operations?

8.9 UNION CARBIDE SAVIN (PESTICIDE) PLANT, BHOPAL, INDIA, APRIL 1988

8.9.1 INTRODUCTION

As stated earlier, the accident at the Bhopal Pesticide plant was the worst chemical plant accident in the world. The plant is located outside of Bhopal, which lies in the center of India more or less on a line through Kolkata (Calcutta) and much to the north of Mumbai (Bombay). The plant was constructed by Union Carbide (UCC) and owned by a number of entities, including the Indian government, and run by Union Carbide India (UCI). The plant was designed to produce Savin pesticide. A material used in the production process was methyl isocyanate (MIC). It was a reaction between water and MIC in a tank that led to this accident. It should be said that the market for Savin was not as prosperous as it once was, and that production was down, which lead to economies in staff and operations being taken. The whole circumstance of the accident and its precursors is obscure, and some attempt will be made here to point out some of the precursors during the discussion about the accident. Given the severity of the accident, one would have expected that everyone concerned would have been anxious to try to find out what actually happened. Unfortunately, the Indian government prevented witnesses from being questioned until some time after the accident, and access to documents was prohibited. All of this went on for a year, until the U.S. courts forced the issue. By this time, the witnesses were dispersed and the memories of the various participants, if located, were different. Also, records had been altered to protect whom? The measures taken to understand the accident have significantly affected the possibility that one will ever know much of the details. Accident investigations are difficult at the best of times, but the delays, distortions, alterations, and loss of memories make this case not only the worst accident, but also one of the most difficult to sort out. The various reports about the accident are uniform in some facts: an accident occurred, it was caused by the explosive release of MIC, the plant was poorly run, and a lot of people got killed in a very nasty way. Even the numbers of people who died shortly after the accident and later are ill defined. One can be sure that both numbers are large and some 2000 to 200,000 have died or been injured. Many reports have focused upon the effects of MIC on people and animals; it is not intended to discuss this here, beyond the statement that the effects are very bad and that MIC attacks the soft tissues, including lungs and eyes, and causes great pain. It appeared that the authorities allowed for the plant to be located close to a population center and that subsequent to plant construction, people were allowed to move into what should have been an exclusion zone. This may not have helped that much, but maybe the numbers of people affected could have been reduced to 100,000.

8.9.2 Description of the Accident

From all of the reports, it is agreed that the cause of the accident was that a significant amount of water entered into one of three tanks containing some 50 tonnes of MIC. During normal operation, the MIC would be cooled by a cooling system. At the time of the accident, this system was not working due to the unavailability of the coolant for the refrigeration system. This may have made the reaction more explosive. How the water got into the tanks is subject to some uncertainty. One concept was that the water entered during a cleaning of pipes, and another was that water was deliberately poured into the tank from a nearby hose via the connection to a removed pressure gauge. The latter suggestion has some credibility, because the working relations between the management and the staff were poor due to the uncertainty that the plant would continue to be operated, given the market for pesticide. According to some reports, UCC was thinking of moving the plant to either Brazil or Indonesia. In this atmosphere of uncertainty, one person was having difficulties with his manager and so was the candidate to be the person suspected of sabotage. Because the independent in-depth investigation was late in being undertaken, it is difficult to prove one way or another, unless someone confesses. A logical discussion about one way that the water may have entered into the tank can be dismissed because of issues associated with the available pressure head from a hose used to wash down some distant pipes. Also, an examination of the pipe leading to the tank was found to be dry, indicating the water did not come from that direction.

The most believable act was that someone did connect a hose close to the tank in order to cause problems. Further, it is thought that the person(s) did not appreciate the extent of the damage that the accident would cause.

The first indication of an accident was a release of MIC and the actions of the plant personnel were to try a number of different things, most of which did not work. The state of the plant was poor due to inadequate maintenance and lack of materials, and this hampered recovery operations. The staff tried to transfer MIC to other tanks, use spray to control the releases to minimize their effect, and initiated alarms to encourage the nearby population to disperse. The plant personnel tried to mitigate the effects of the accident, but in the end, they had to escape themselves. The plant was badly designed and poorly run, and there were inadequate safety precautions, such as better containment of dangerous materials, improved sprays, increased dump transfer capacity, and better warning and evacuation procedures for the nearby public. It is interesting that UCC and UCI did not have sufficient money to run the plant properly but were able later to pay out $470 million (court settlement) plus $118.8 million in other monies. This should be an indication to upper management to pay attention to operational practices and design details.

One cultural issue that remains to be discussed is a limitation in the engineering culture. Engineers seem to design things and assume that their designs are going to work properly. They also need to think about what if they do not. Application of HAZOP can be very useful for "what if" situations. In this case, if the combination of water and MIC had been thought of as leading to an accident of this magnitude, then several design features should have been added. Even the idea that the only way for water to get into the tank was by sabotage and every other way was impossible indicates the mind-set of the design and operational personnel, and also the investigators. The protection systems should have been working on the basis that things can go wrong, especially with substances like MIC. The words of Lord Cullen mentioned in the case of the Piper Alpha accident come to mind relative to the use of nuclear plant design considerations as far as safety of plant is concerned.

Figure 8.14 shows the event sequence for this accident. The figure shows a number of features that should have helped the personnel prevent or mitigate the effect of the action taken to introduce water into the MIC tank. One problem is that the defense methods did not work properly; therefore the end states for many of the branches are the same with or without the operations being taken. The figure is used to emphasize the fact that operator actions could have affected the accident consequences and that the plant personnel tried to do something to change the course of the accident. For the figure, the following assumptions have been made:

1. The alarms work and increase the reliability of operator actions.
2. Transfer works well, but there might be a delay and some MIC escapes.
3. Manual spray is designed well and is fairly effective, but not as effective as the transfer.
4. Public warnings are given, the public is informed and practiced, but only 80% to 90% obey the instructions.

It is difficult to use descriptive expressions to show the differences between the various branches. One really would need to define the ratio of deaths prevented by use of the various suppression and control systems.

Introduction of water into MIC tank	Alarms sound	Manual action to transfer MIC	Manual water suppression	Public warning to evacuate	Disaster

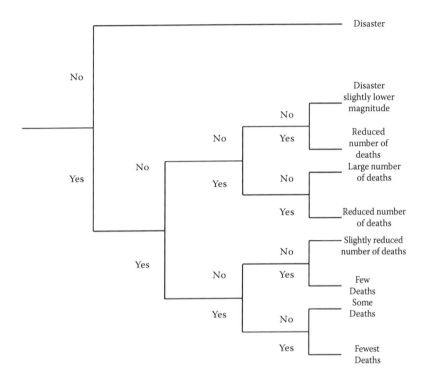

FIGURE 8.14   Event sequence for the Bhopal accident.

The net effect for a system design is that the responses are not entirely successful in the prevention of an accident or deaths. The prevention of all escapes of dangerous chemicals needs a complete change in the design and operational aspects of chemical plants. Also, it might need the chemical industry to be regulated more like the nuclear industry and to have a trade organization the equivalent of the Institute of Nuclear Power Operations (INPO).

The Bhopal plant needed a number of improvements in design and operation in order to prevent the release of MIC. It appears from descriptions of the actions taken by the staff at the plant that they did their best to try to minimize the effects of the accident. The figure shows that a number of features either did not work or did not work well enough to help the staff in their endeavors. From the reports, it is not clear that the alarms worked, but the staff did respond to the effects of MIC. They tried to transfer MIC to another tank and did try to use sprays to control the release or the extent of the release. The warning for the city dwellers was sounded; unfortunately, the warning was terminated by order to prevent the panic of the townspeople, so it was better to let them die rather than warn them—very strange logic. It would appear that beyond the personnel problems at the plant, the staff did take actions to try to control the effects of the incident. The message that one gets from this review of the accident is the irresponsibility of the management to maintain the plant in good working condition. If it did pay to operate the plant efficiently, then the decision should have been taken to shut the plant down. The question about Indian government politics enters into that decision, and this may have been the reason for not shutting down the plant. It does seem to be suspicious that the government controlled everything after the accident for at least a year.

8.9.3 ANALYSIS

1. Management/personnel relations were not very good at the plant.
2. Management was not very responsible toward running an efficient plant by ensuring the plant was well maintained, such as ensuring that safety systems worked, that personnel were provided with good training and procedures, that evacuation alarms worked, and that the processes were practiced and involved Bhopal citizens.
3. Despite the problems at the plant, the staff tried their best within the limitations of the plant and its systems.
4. It appears that the most likely cause of the accident was the disgruntlement of the staff and one person in particular. It is likely that he did not appreciate the extent of his action.
5. Given the plant condition, plant manning should have been more carefully controlled, so that the one miscreant was not in a position to cause so much harm. There is enough history to know that sabotage can occur under poor industrial situations. Management needs to ensure that if it does occur, it is only a nuisance and does not result in a catastrophe.
6. The management had been warned during previous visits of the poor safety configuration of the plant, for example, the MIC tanks contained large quantities of MIC, and it would be better to have a larger number of small capacity tanks. If a good PRA/HRA safety analysis of the plant had been carried out and the management had paid attention to the findings, issues associated with MIC, personnel controls, confinement, effective sprays, and practiced evacuation of townspeople would have meant that the accident would not have been so severe if it even occurred.

8.10 FLAUJAC: EXPRESS/LOCAL TRAIN CRASH, AUGUST 1985

8.10.1 DESCRIPTION OF THE ACCIDENT

The crash at Flaujac was between an express rail car going from Brive to Capdenac and the Paris-Rodez passenger train on the same single track going in opposite directions. The details of this accident are given in a book by Carnino, Nicolet, and Wanner (1990). The book contains details about the accident, but this type of accident occurs fairly frequently in the rail system. Just recently a similar accident took place in Los Angeles on Friday, September 12 (*LA Times,* 2008). Although the reports of the L.A. crash are there, the details of the crash still need to be analyzed by the National Transport Safety Board (NTSB). The blame for the accident has been as due to the driver, who died in the accident, but this is often the first thing pointed out.

The French accident took place in the French Department County, called LOT, in southwest France near the Flaujac railway station. The collision took place at about a 200 km/hour closing speed. There were 31 people killed and 91 injured. Subsequently, oil carried in the rail car leaked and caught fire, causing more destruction.

The Brive-Figeac section of the railway is 90 km of single track with some six crossing points including the Gramat and Assier stations, where the tracks double so that an express train can pass a stationary train. The intent on the day of the accident was for the two trains to pass safely. The train schedules are selected to match the safety needs, and the station masters are informed of the schedules and kept up to date. Trains, however, can run behind schedule, and the process has to be flexible enough to accommodate delays. There are a set of regulations that station masters have to follow to ensure train safety. The rules have to be followed and are essentially designed to capture passenger trains moving in one direction at what could be called a lay-by or station halt.

It turned out that the passenger train (6153) left Brive late (15 minutes), arrived at Gramat late (13 minutes) and then left for Assier at 15:40 (1.5 minutes late). The rail car (7924) left Assier one and one-half late. Therefore they were both on the single track. According to the schedule for this time of the year, the 6153 train (service June to September) was due to be passed at Assier. For the replacement passenger train service (6151), service was from September to June and the cross-over point was Gramat. The station master was confused as to which schedule he was using, and he was helped in his confusion by a conductor of a crossing at Gramat rather than Assier. After some moments, the agent realized his mistake and called the other agent, but the communications were not clear and the railcar was sent on its way to Gramat. The collision took place at Flaujac between Gramat and Assier.

The passenger train should not have been at Gramat until the express train had passed it. The trains crashed at 15:48, some 8 minutes after the passenger train left Assier. So what are the ingredients that led to the accident? Clearly, the coordination of the passing times is crucial. The signals controlling the passenger train have to operate to prevent the train from leaving too soon, so as to be on the same track as the express train. It sounds simple, but it requires communications between stations tracking the trains and being aware of what train is what and its scheduled time at any given stretch of the track. The elements for confusion exist: Were the communications accurate, were they correct in terms of identification, and were they timely? Are the requirements to hold a train understandable to the person controlling the signals, and are the signals properly identified by the train driver? What appears to be a very simple and straightforward concept starts to be less clear. If the train schedule is varied from month to month, the station staff may be confused as to whether the train is late or not, or even if the train is the 7924 as opposed to the 7921.

It appears from the records that there was a misunderstanding about train numbers, and furthermore, the strict rules relative to stating the number of the train that

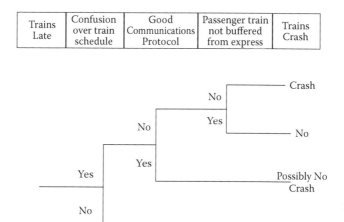

Trains Late	Confusion over train schedule	Good Communications Protocol	Passenger train not buffered from express	Trains Crash

FIGURE 8.15 Event tree showing the various elements compounding the accident.

the agent was passing from one station to another were not clear. If the communications were clear to both parties, there could be cross checking one with the other. Also, a contributing factor was the fact that both trains were running behind time. Any delays in the railway system can lead to delays percolating through the system. This can lead to unreliable and unsafe operations.

Figure 8.15 shows some of the elements that either led to the accident or compounded it. If there was no confusion over the timetable and the train numbers, it is unlikely that an accident would have occurred. The poor communications may have meant that the likelihood that a recovery by the station staff was not possible. However, the recovery depended on the agent becoming aware of his error early enough. If the passenger train was contacted and took refuge in a bypass, then it would be buffered from the express railcar. There is a station at Flaujac, but the question becomes, is there a bypass line there and can the warning get to the passenger train driver in time? Included is the possibility of another bypass beyond Assier and Gramat. Failing to lead the passenger train into a bypass, the accident is certain. It happened on August 3, 1985, and 31 persons were killed and 91 injured.

8.10.2 ANALYSIS

1. Single-line railways present a danger, because for trains to be safe, they have to cross at sections consisting of two tracks. They require close attention by rail personnel. Rules and the schedule govern which train is on a single-track zone, and the staff needs to know where the trains are at all times. They also need to prevent the entry of another train into a single-track zone where there is another train. The aim is to ensure that the express train passes the passenger train, while it is waiting in a bypass (station). Signals are provided to control the movements of the trains.

2. The use of similar numbers to identify different trains can cause confusion, especially if similar train numbers cover effectively the same train but are running at a slightly different time or season. The case in point are the passenger trains 6153 and 6151—they are effectively the same train going to Paris from Rodez, except that the schedules are slightly different. For one train (6151), the cross-over section with the express railcar (7924) was Gramat, and for the other train (6153) it was Assier.
3. The situation was made worse by the fact that the trains were both late. This added to the confusion.
4. Communications between two agents were poor, and they did not use a repeat-back protocol. Even the mention of the affected train's number was not stated during the emergency call made by the agent who became aware that there was a problem. Reports on whether direct communications between trains and agents were available and working were not known. If they had been, the trains could have been warned to stop mid-zone before the accident occurred, even if it meant that one of the trains had to back up.
5. The last issue was that one of the agents was a reserve agent and may not have had the same experience as the regular agent. Mostly, the trains do not have accidents.

8.11   UNDERGROUND FIRE AT KING'S CROSS, MAY 1990

8.11.1   Introduction

A number of accidents have occurred in the rail systems throughout the world. One accident occurred in an escalator area of the King's Cross underground system. The Underground, as it is called in England, has been operating for a number of years, and this accident brings up the fact that accidents can occur even after many years of operation and can even reveal a new phenomenon. Another point that could be made is that maybe the organizational vigilance possibly had diminished with time? This often appears to be a contributor in many accidents. The author came up with a 10-year rule that organizations appeared to use: "If an accident has not appeared in 10 years, then everything is fine." Of course, he was being sarcastic.

The London Underground system has grown over many years, and each phase has seen the move to a multilevel, interactive system. It started in 1885 as a modified surface system, but in time, the system has grown, and the tunnels have become deeper. Also, because of the topology of London, a basin with the Thames running through the middle, one sees cases where the station ticket entrances are even at the same height as the top of St. Paul's Cathedral, which is in the middle of London. In general, railways are designed to be flat; too much of an incline would produce an inefficient drive system. The trains are basically of the steel wheels and steel track type.

The King's Cross and St. Pancras Stations are a complex set of linked underground railways carried in shallow to deep tunnels. There are a series of links between the rail systems, some by tunnels and others by escalators. At King's

Cross, there are six lines coming together: Northern, Piccadilly, Circle, Victoria, Hammersmith and City, and Metropolitan lines. Some of the lines are at the surface level and others are deep, like the Piccadilly line. The various levels reflect the development of the Underground system over the years. Replacement of equipment has been gradual, for example, on the escalators on the Piccadilly line, the steps were wooden. These wooden escalator "steps" played a key role in the King's Cross accident.

8.11.2 DESCRIPTION OF THE ACCIDENT

The fire broke out at about 19:30 on November 18, 1987, on the escalator of the Piccadilly line. The Piccadilly line is deep and the escalator is quite long. It was fortunate that it occurred after the evening rush, otherwise many more people might have gotten killed or injured. Like most accidents, there were precursor indications of trouble; fires have broken out at earlier times in the Underground. There was a fire at the Oxford Circus station. As a result, smoking was banned on the Underground as of 1985. However, after the King's Cross accident, investigation showed that persons, on their way out, used matches to light up cigarettes and cast their matches down on the escalators. These matches could fall onto flammable materials under the escalators.

It has been concluded that the majority of matches did not cause fires, but some did cause some burning as determined during the postaccident investigation.

In the case of the King's Cross accident, the mixture of lubrication grease and fibrous materials caught fire, and then the fire spread to the wooden steps. At the beginning, the fire produced fairly clean white smoke from the burning of the wood (steps and sides); later there were to be flames and intense smoke. Initially, the firemen were confident that it could be handled, because they had dealt with some 400 such tube fires over a number of years. The precautions to evacuate the passengers taken at this time were little because of the belief that this was a controllable event. Unfortunately, this was not to be case, and because of train movements, there was a pumping motion of air that led to large air flows (this is a phenomenon well known to tube travelers), and this may have added to the risk. There was a flash situation leading to filling ticket halls with smoke and flames. As the fire progressed, even the paint on the walls caught fire under the intense heat. It has been said that as a result of the paint burning, the smoke went from thin and white to black and oily.

Not only was the wood burning, but there was a mixture of grease (from greasing the escalator mechanisms), fluff, paper fragments, and hair from various sources. The grease was of the heavy variety, slow to ignite but once burning supplied a lot of energy, and there was a lot of grease accumulated over some 40 years.

There appeared to be new phenomenon involved called the "trench" effect. The walls and tread of the escalator acted as a trench to direct and intensify the fire and its effects. Interesting was the fact that a subsequent analysis carried out by Oxford University showed this effect and the 30 degree rise angle of the escalator played a key role. This analysis was not believed until a real model of the situation duplicated the same effect. It was also believed that fires do not burn downward, whereas there are circumstances when a fire does advance downward because of radiation effects and particle transfer. Another thing observed was that the right-hand side of the escalator was the side that showed signs of burning. Standing passengers are most likely to light up, despite the rule not to. As a result of the fire, and a lack of an understanding of the potential severity of the fire at the beginning, the fire grew and became much worse, resulting in 31 persons dying. In addition to those who were killed, 60 people received injuries ranging from severe burns to smoke inhalation.

A public inquiry was held into the incident and was conducted by Mr. Desmond Fennel, QC. It opened on February 1, 1988, and closed on June 24, after hearing evidence for 91 days.

Figure 8.16 presents the event sequence for the accident and indicates that a combination of steps needs to be taken to improve fire safety aspects, including passengers following rules, and the organization updating equipment to minimize the spread of fires and removing sources of fire hazards in the underground environment. The lower leg of the figure is for the case where the public follows the rules

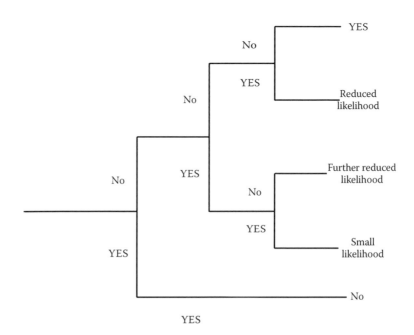

FIGURE 8.16 Event sequence diagram for the King's Cross fire.

without exception. This is unlikely to happen, so it is still necessary to ensure the other actions are taken to drive the risk probability to a very low value.

8.11.3 ANALYSIS

1. It appeared that the London Transport Underground did not appear to be aware of the disastrous consequences of a big fire in the underground. The presence of heavy lubrication grease and fibrous materials under the escalators as a source of combustible material seemed to be something passed by for a number of years. The buildup of grease took place over a number of years. Tests indicated that the grease was hard to ignite, but the combination of grease and fibrous materials was more capable of burning.

2. The trench effect and the impact of the 30 degree slope as part of the fire phenomenon was unknown before this fire. One could not be expected to predict these phenomena. However, the idea that fires do not travel downward should be discarded, because of radiation effects and the fact that gravity plays a part in the interactions.

3. The management of the Underground did not use risk-based methods to evaluate the potential for fires. They could have used PSA or even HAZOP techniques to alert them to the potential impact of a combustion source associated with the escalators including the wooden escalator treads and sides. They had initiated a no-smoking rule, but it does not seem to have been vigorously enforced.

4. The layout arrangement of the passenger access tunnels have just grown in number. More thought is necessary to ensure that the passengers could evacuate quickly and without high risk. This goes for all manner of risk as well as fires.

9 Analysis of Accidents Using HRA Techniques

INTRODUCTION

In Chapter 8, eleven accidents were described that have occurred in different industries (see the list in Table 9.1). There are lessons to be learned from these accidents; most involve the following elements: equipment failures or limitations, human errors of one kind or another, and the impact of management decision making on the accident.

In this chapter, we apply some human reliability methods to these accidents as though the accidents have not occurred a prospective sense. One may try to determine what the overall risk factors covering each type of accident would be. While it is not possible here to undertake a complete PSA for each of these situations, it is useful to examine the accident occurrences from an HRA point of view.

The accident analyses show that management decision making is an integral part of accident progression; therefore this influence needs to be incorporated into the PSA/HRA analysis. The term "organizational factors" has been developed to cover the impact of management decision making in risk assessment. The idea of organizational factors means that it is difficult to separate the actions of the operators from those of management. The sharp ends of accident responses are carried out by operators, who take the actions either rightly or wrongly. However, management makes decisions that influence the actions of the operators. The word "operator" covers all aspects of operations by personnel from maintenance and test to controlling power levels at power plants to flying aircraft. Operators are in charge of operating machinery, and they make tactical decisions to terminate or mitigate accidents.

So what does management do? They do not operate the equipment, but they determine the environment in which the operators function. Many of their decisions are taken under advisement from experts; for example, they decide to buy a power plant of one kind or another, which itself may contain predesigned items. Although they are responsible for the decision to buy a particular design, the design may contain items that have not been explicitly decided by management. The manufacturer plays a major part in the final power design; in fact, the majority of the plant features are the outcomes of their design process. Part of the decision process undertaken by management is the examination of alternative power plant designs and selection of which plant best meets their requirements. Although there are some choice limitations, management plays a significant role in the process. Management usually focuses on the main features of cost, operational flexibility, and safety. In the case

---

TABLE 9.1

List of Accidents Considered for
Human Reliability Assessment

Name of Accident

Three Mile Island Unit 2—Nuclear accident

Chernobyl—Nuclear accident

NASA Challenger Orbiter accident—Space

NASA Columbia Orbiter accident—Space

Tenerife accident—Airport

JFK Airbus 300 accident—Aircraft

BP Texas City accident—Petrochemical

Piper Alpha accident—Oil rig in North Sea

Bhopal pesticide accident—Chemical plant

Flaujac train accident—Railway

King's Cross Underground accident—Railway

---

of nuclear plants, management is deeply aware of regulatory requirements in order to ensure the USNRC will ultimately license the plant. The relationship between the plant management and the licensing authority will not be discussed here except to say that the licensing authority plays a key role and can deeply affect cost and availability of electricity to the user. Most important, the safety of the plant and its operation are of prime importance to all parties. The licensing and operation of nuclear plants are very heavily impacted by safety concerns; other commercial operations may not be as heavily affected as is the nuclear industry.

Some details of plant design are more important than others to the decision makers, so, for example, the price of a power plant may be more important than the layout of the control room. The manufacturer will usually work with the power plant owners on the detailed design of many features; for example, variations in the basic control room layout, selection of panel instruments, and so forth.

Management decisions lead to the environment in which the operators find themselves. This is true for other things that affect the ability of operators to perform, such as the training environment and quality, the layout of the control room equipment, and the procedures.

For other industries, management decisions can include the selection of specific equipment. In the case of NASA, one is really covering very limited production runs, such as four or five shuttles, or even one International Space Station (ISS). Here NASA management is central to the design process, construction, operation, and maintenance. The U.S. government provides the budget that controls what NASA can do. Other nations, such as Russia, provide capsules and modules, so NASA is not in complete control of the ISS design. In fact, there are different design concepts and equipment involved in the ISS vehicle. The ISS vehicle is an international project involving many different nations.

9.1 INVESTIGATION OF CHAPTER 8 ACCIDENT SITUATIONS

In this section, I set out the approach to performing an HRA analysis of an accident situation in a prospective manner. What does this mean? Clearly determining the probability of accidents is one approach, as one might be interested in understanding what the estimated probability might be if one had been performing a PSA/HRA before an accident had taken place. In other words, would the expected probability of failure be high? This is what is meant by a prospective analysis. The main focus is on the human reliability aspects of the analysis.

It is always worthwhile examining theories and techniques to see if they work and have face validity. By performing a prospective analysis, one can gain some insights into the utility of a specific model and the associated thinking. An event that occurred quite a while ago encouraged me to carry out such a study. A PRA had been performed on a BWR plant situated in New York state. The core damage prediction results placed the performance of the plant in the middle of all BWR plant results. One of the senior USNRC managers said that he could not understand why the plant results were as good as that, because the plant was on the USNRC watch list for underperforming plants, so he expected that this would show in the results. If the operational characteristics mirrored the watch list, then one would expect that the estimated core damage frequency would increase for "poor" plants and decrease for "good" plants. The implication drawn by the USNRC manager was that the methods used did not capture the real operational characteristics of the plant, an interesting observation with some merit. The question remains: Was the lack of difference due to the fact that the items used in the URNRC watch list evaluation were inconsequential as far as risk was concerned, or did the methods used in the PRA not capture their significance?

Another situation like this occurred in the case of the PRA performed for the ISS. The PRA was reviewed by a high-level committee, and among the various comments made was that the HRA method used (Hollnagel's CREAM I method) did not account for NASA experience or knowledge. The issue raised by both of these cases, and there are surely others, concerns whether the use of HRA methods based purely upon generic data properly addresses the risk of operating an actual system. The risk might be bigger or smaller, and it should be better defined, so that one can determine where the risk is and what can be done about it. In the case of the ISS study, steps were taken to capture NASA knowledge and experience about the ISS operations. These insights were incorporated into the study. The HRA method used was the holistic decision tree approach. Chapter 10 covers main elements of the ISS study as an example of an application of an HRA method to a real situation. This method allows for various factors that influence human reliability to be accounted for, such as procedures and human–system interface effects. Furthermore, the quality of each of these factors can be adjusted to cover human factors influences and management decisions.

The general organizational approach will be to try to follow the normal HRA approach based upon a modified Systematic Human Action Reliability (SHARP) method. HRA investigations are usually carried out by examining plant operating records, training operations, and simulator training exercise records and by interviewing plant personnel before commencing the actual HRA task. For the HRA analyses covered in this section, the author had to rely upon information that was available within the accident reports, which maybe very much less than one would like. Therefore, some of the data had to be assumed by the author in order to move ahead.

The organization of HRA studies was covered in Chapter 3, and the reader should refer to Figure 3.5. As to the choice of HRA model, this will depend on the review of any specific accident and the assessment of which HRA model may be suitable for a given situation. It could be that a number of models or methods could be used. Under these circumstances, one should pick one just to illustrate its use. As previously mentioned, the three main contributions to performing a good HRA study are the skills of the HRA analyst, the relevance of the HRA model to the situation, and the availability of HRA data relevant to the situation. The question arises as to the availability of reasonable data for every case. It is not clear that relevant data are available. Chapter 12 contains a discussion about databases and data banks. A detailed examination of databases is not carried out here. The corresponding human error probabilities (HEPs) are estimated here. It is anticipated that some HRA experts deeply involved in specific fields might disagree with the numbers based on their experience. The aim is not to try to establish necessarily highly accurate HEPs, if this was possible, but to obtain numbers to close the loop in the prospective analysis. In cases where it is not possible to perform a detailed HRA study to determine an HEP, the author will make an estimate.

Nine accidents have been analyzed (Table 8.1). Each of the nine accidents is discussed here from an HRA point of view, starting with the TMI Unit 2 accident. The depth of analysis given in Chapter 8 for each accident may not be the same, so there may be gaps in the HRA analyses. Some of gaps in the accident analyses are due in part to the nature of the accident and to the availability of the official and unofficial results. In some cases, the author followed the reporting of the accidents fairly closely and other cases not so closely. For example, the TMI accident very much affected the industry in which the author was working. When I was at General Atomics, a couple of us were requested to make a presentation to the management of Royal Dutch Shell (one of GA's owners) on the accident and its consequence for the nuclear industry. Later, at the Electric Power Research Institute (EPRI), the TMI Unit 2 accident was analyzed in depth, and the author witnessed much of the analyses and its results. Lastly, the author was involved with others in the estimate of the survivability of containment instrumentation at TMI following the accident. This required an understanding of the core damage and its effects.

As to the other accidents, the intent was to try to understand the accident progression, the influences that led to the accident, and the subsequent effects. The coverage of each of the accidents is not to the same level. As always, Wikipedia is a good starting source for a discussion of an accident, but to be sure of the facts, one needs to dig deeper. In some cases, reports have appeared on the Web that can be identified through Google. Some organizations release their reports, and others do not. As stated earlier, the quality of both the reports and the investigations are not necessarily to the same standard. For example, the Piper Alpha accident was analyzed and reported in depth in a British government inquiry and led to a two-volume report on the accident by Lord Cullen. It was a detailed investigation that involved a large number of witnesses and experts. The reporting of the Piper Alpha accident in Chapter 8 does not compare with the above-mentioned report by Cullen. One can only hope that a reasonable report on the accident and the main influences have been covered. Other reports on accidents do not reach the level of the Cullen report. These in-depth reports do take a while. In the case of aircraft crashes, the debris has to be assembled before the causality can be worked out. Even in these cases, it is not possible to divine the truth. In the analysis sections of the accident reports in this book, an attempt is made to address some of the causality aspects. Event sequences have been drawn as an aid to the reader to understand the main influences that led to the accident.

9.2 ANALYSIS FROM HRA POINT OF VIEW

In each analysis, one needs to understand what influences could play a part in the possibility of an accident. Normally, in a PSA, we look at a number of potential initiating events so that we get an overall understanding of the integrated risk and what sequences are more risky than others. The initiating events that lead to bigger contributions to risk are then called the dominant risk contributors. In this analysis, a given initiating event has occurred, but the object is to examine the situation as though there was only one initiating event, the one associated with the accident.

Typically in a PSA, one would construct an event tree and associated fault trees and then start to populate the trees with estimates of initiating event frequencies, equipment failure probabilities, and HEPs. Also, as part of the PRA/HRA process, data associated with running the plant; historical data related to equipment failures and plant incidents; simulator records; reviews of training programs for operators, maintainers, and testers; procedures and their current status; and control room equipment and layouts, including details about computer systems, would be collected. In other words, the PSA investigators would perform a thorough review of the plant, personnel, and operating history, including interviews with management and staff. In the case of the analyses considered here, all of the relevant data have to be extracted from the accident reports together with some estimates based upon the author's experience.

In each case, estimates for both hardware and human-related numbers must be estimated. The event sequence diagrams (ESDs) associated with the accidents give one a basis on which to start the analysis. The description of each accident can enlighten us to some degree as to what the environment is as far as the operators are concerned.

As much information as possible should be extracted. The following list can help organize the analyst's review of an accident.

1. Management role in the accident: Does local management have to defer to distant management before taking action?
2. Experience and training of the plant personnel: operational and maintenance and test and calibration.
3. Maintenance records, equipment failures.
4. Design features of control rooms or control panels.

5. Design decisions taken in the past, if covered.
6. Manning schedules, reduced or enhanced.
7. Decision making, distributed, or concentrated: Who is the responsible party for making decisions?
8. Role, if any, of regulators.
9. What is the environment under which the accident took place? Is this different from normal?

The process is to try to use as much information from the accident related to the influences that could condition the responses of the personnel. Like this, one can capture all of the extractable information from an accident report. In some cases, the report focuses on those things that support a deep investigation of factors that affect the human. Other reports are less focused on these things; therefore there is more uncertainty about the results. The viewpoints of accident investigators are not the same as those of human reliability analysts. The accident investigators look for front-line causes of an accident, and HRA analysts try to predict the probability of systematic actions rather than the probability of an action of a single individual.

The level of detail relative to human performance varies from one accident to another. The object here is to try to understand the context under which these accidents take place and see, based on this information, if one can deduce what the predicted HEP would be. Here the HRA analyst is not in control of the accident investigation. However, these accidents are real, and the contextual relationships are there to be studied. The process of the HRA is a good way of seeing various HRA models in real situations and their values.

Say that you collected information related to the items listed above. The question arises as to how this information relates to the need to estimate HEPs. The information helps the analyst to select an HRA model, which should be related to the specific situation. Some of the information speaks to the task being undertaken by the crew, but it also covers the context in which the crew is trying to respond to the accident. An analyst's experience in attempting to analyze the situation helps him or her decide what the key influences are and then select an HRA model that best represents these influences. This process works for both a conventional PSA and for this analysis.

The analyst also has to consider the data requirements of any model. Data sources may be available from plant records, including training records derived from simulator sessions. Another possibility is to use expert judgment. In accident studies, much of this information may be available but not necessarily recorded as relevant.

9.3 THREE MILE ISLAND, NUCLEAR ACCIDENT

The details about this accident are covered in Chapter 8. Many of the key parameters related to the accident are given in Table 9.2. The first column of the table covers the key items. The second column covers the evaluated status of these items with respect to the accident. The third column deals with the impact of the items in determining the contribution of the items to the accident, in particular with regard to human reliability. This table reflects the list of contributors covered in the previous section.

TABLE 9.2

Key Human Factors Elements Associated with TMI Unit 2 Accident

Items	Status	Assessment of Impact
Quality of plant management	Typical for 1970s, failed to ensure operations and maintenance rules were emphasized	Failure to understand plant risk, not really exposed to the use of PSA techniques
Quality of crew skills (assessment)	Average to good for then-current requirements	U.S. Nuclear Regulatory Commission finding prior to accident implied that they were among the best in the industry
Quality of training: classroom training, team training, communications, and simulator	Inadequate for situation	Limited exposure to simulator training, no multifailure transients
Simulator	Inadequate simulator availability, limited simulator capabilities, not really full scope	Very few simulators were available, most simulators were limited in terms of what transients could be run; the U.S. Nuclear Regulatory Commission later required every plant have a full scope simulator
Quality of procedures	Poor for situation	Subsequent to accident, procedures moved from event based to symptom based
Quality of crew knowledge	Below need, to focus on pressurizer "solid" state; limited knowledge of pressurized water reactor dynamics	Reduces probability of determining problem/solution
Quality of human–system interface (Control Board)	Industry average; needed to be improved	Does not help crew's focus on key effects; later U.S. Nuclear Regulatory Commission requested human factors review of all main control rooms
Design decisions related to PORV operation	Poor	Pressure operated relief valves gave no direct indication of closure, only of the demand control signal
Control of maintenance activities	Inadequate	Failure to ensure labels did not obscure auxiliary feed isolation valve status
Role of regulatory authority	U.S. Nuclear Regulatory Commission underestimated the importance of the crew in responding to accidents	The U.S. Nuclear Regulatory Commission postaccident response was mainly focused on human factors and reliability improvements

Clearly, the detail with which one can determine the contributors depends on the depth of the investigative reports and conclusions drawn by the investigators. To some extent, the author has taken the privilege of drawing the conclusions, because the main purpose of these investigations is to point out to analysts what the depth of their HRA investigations should be.

9.3.1 ANALYSIS OF THE ACCIDENT

As seen in the ESD (Figure 8.7), there are a number of events of importance or interest in that they affect the accident progression: loss of main feed due to failure of the switching operation, failure of auxiliary feed water to start, PORV fails to close and reactor pressure dropping to below set-point, PORV design fault to directly indicate PORV open or closed. The consequence of this was that the primary pressure dropped, boiling occurred within the reactor core, and a void was formed in the top plenum forcing water out of the reactor vessel and into the pressurizer. The level in the pressurizer rose to the point that the crew thought that the pressurizer would go solid. They then thought there was enough water covering the core and it was safe to shut off the safety injection (SI) pumps. Later, due to new information (supervisor from Unit 1 informed them he thought that they had a small break LOCA), the crew turned the SI pumps back on and the hot fuel elements collapsed due to cold water impingement.

Failure to have auxiliary feed water meant that heat-up of the primary side was faster, because cooling of the primary side by the steam generators (SGs) was limited to the availability of stored water in the once-through SGs (OTSG). For OTSGs of Babcock and Wilcox design, the water storage is small and there are about 15 seconds of full-power steaming from the stored water without replenishment, so the crew does not have much time. (The figure for a Westinghouse PWR is about 15 minutes.) The typical response to a reactor trip is for the primary pressure to rise, which would lead to the PORV lifting anyway. The loss of main feed leads to a reactor trip.

The reliability of the PORVs is not very high, but in this case, it may have been even lower. An estimate of the failure to reseat is around 0.1 per event. The number of events was, at this time, about two to three reactor trips per year. One could look at the probability that both auxiliary feed-water valves were isolated as about 1.0E-02 per maintenance/test operation with about two such operations per year. Now the crew probability of isolating the SI is more complex, because it is a function of training, alarms, and so forth. For this case, the HDT method will be used. Clearly, if one carried out the analysis at the time, even the assumptions would be different, because the perception of the quality of the various influences by the analysts would be different than now.

A number of persons would consider that this accident was categorized as an error of commission (EOC), and it is, if the operators were PRA analysts, but they are not. The crew thought that they were taking the correct actions according to their training. It is an error induced by the training undergone by the crews and is likely to be a systematic rather than an individual failure. One just applies the same rules as for any situation rather than trying to differentiate between errors of commission or omission. Because the design of the MMI has changed, the training process has changed, and the exposure to complex transients on the simulator increased the awareness of the limitations of the old MMI, training, and so forth. One's ranking of influence effects could be affected. Therefore, two estimates will be given of the HEPs based on 1979 and 2009 rankings.

The form of HDT to be used in this case is that there are four influence factors (IFs) of equal weighting (0.25), and the assumed anchor values are 1.0E-03 and 1.0. The quality factors (QFs) for each of the IFs are given in Table 9.3 for the two cases. For an understanding of the interrelationships between the various factors (IFs, QFs, and HEPs), see Chapters 5 and 10. The two cases covered are estimates as of 1979 and 2009. The output from the HDT Excel program gives the estimated HEPs. The range of the QFs is 1.0, 5.0, and 10.0, and these represent good, nominal, and poor qualities. In applications of HDT, the definitions of each IF and QF are given in detail. The definition of the MMI is centered on the various changes in human factors. It is felt that one can refer to HFs experts to be able to relate the design of the MMI, which should follow HF guidelines, to the intrinsic quality of the interface. One should be capable of doing this in light of all of the experience obtained in the design and operation of MMI over the years. It is interesting to note that the field of usability started to fill the gap that can potentially exist for visual displays as far as acceptance and functionality are concerned. Tests have been carried out to test emergency operating procedures (EOPs) in much the same manner, using simulators to ensure that they are both accurate and usable. More work needs to be done to codify, if possible, the lessons learned from these exercises so one can begin to understand what is poor, nominal, and good in the field of human–system interfaces (HSIs). The same goes for defining the aspects of the quality of crew training and procedures. In this book, the author will make the estimates for the TMI Unit 2 accident for the purposes of these analyses.

As stated above, two cases are computed for TMI Unit 2 human actions related to SI being switched off. The first case is for the period before the TMI accident, and the other is much later, say 2009. In HDT analysis, three ranges were selected for the MMI quality. For the first period, a value of QF for the MMI is selected to be 1; the quality of the MMI is judged to be good by the standards of the time. Over time and especially after the TMI accident, it was realized that the MMI was not that good, and now the QF value is assessed to be 5 (i.e., neither good nor bad). In some QFs,

TABLE 9.3

Assessment of Human Error Probabilities for Three Mile Island Unit 2 Crew Safety Injection Actions

Item	QFs Old Assessment	QFs New Assessment
Man–machine interface	1 good	5 nominal
Training	1 good	10 poor
Experience	1 good	5 nominal
Procedures	1 good	10 poor
Human error probability	1.0E-03	1.47E-01

the change is from good to poor (e.g., the training of the crews by today's light is judged as being poor and undoubtedly the feature that contributes most to accidents and their consequences). Similarly the event-based procedures are to blame. One notes that the procedures, training, and MMIs have been changed worldwide as a consequence of this accident.

Going through the assessments, it appears in 1979 that the HEP would have been given the value of 1.0E-03, which would be thought to be an acceptable number. However, if one looked at the situation in the après accident environment, the assessment for the TMI operation would be nearer to 1.5E-01. This would have been an unacceptable answer, especially if taken with an initiator frequency of 2 to 3 per year and a failure of the PORV operation of 0.1. The core damage frequency (CDF) for this event would be estimated to be 1.5E-01 × 0.1 × 2 = 3.0E-02/year. This does not compare very well with a maximum of 1.0E-04 for the CDF. Even the old assessment yields a figure of 2E-04 for the CDF.

Now one would be looking to drive the HEP to at least 1.0E-03 and keep it there. The PORV system needed to be improved so that if the value stuck open, the information would be relayed to the crew and they would isolate the failed PORV. The joint probability of the PORV failing open and the crew failing to isolate the valve would be less than 1.0E-01 × 1.0E-01 = 1.0E-02. Now with a good MMI design, the probability of the crew failing to isolate would drop to 1.0E-02; therefore, the joint probability would go to 1.0E-03. Under these circumstances, the CDF would go to 3×E-3/year. If training and procedures and MMI are improved, then the numbers can greatly improve. With improved operational care of the plant, systems, and components, the number of trips (or test of the system) decreases to less than 1.5 trips per year, so the CDF from this type of event drops by even more.

The above calculations are speculative, and they are the author's estimates, but the message is clear—the accident is made up of a number of components that have to be considered. The designer of the plant and the management team make decisions on pieces of equipment, possibility based on cost, that can turn around and hurt the safe operation of the plant. But it is difficult to see how the acquisition of this type of valve could lead to this accident, or how the failure to train the operators on the dynamics of the plant could lead to an accident, or how the design layout of the MMI could contribute to the worst U.S. NPP accident. One could say that this might make the case for building a simulator well before the plant is constructed to test some of these things. Just having a transient plant code is insufficient for this purpose—one needs to involve operators.

The worst combination of the figures indicates that the accident was fairly predictable to have occurred within 30 years, but the event might occur more quickly if the value always failed to shut, after lifting. In this case, the CDF could occur within 3 years. If the other QFs were adjudged to be worse, the period for a CDF would continue to drop. For example, the experience of the crews was taken as nominal, perhaps in this case it was poor, probably because simulator training was limited and most likely inadequate as far as this incident is concerned. If one continues on in this manner, the accident becomes inevitable, and perhaps it was. But the other message is to be very conservative in estimating HEPs and to see if there is any evidence to prove or disprove such conservative estimates, like the examination of simulator sessions.

9.3.2 SUMMARY

This accident was the worst U.S. nuclear-related accident, at least in terms of cost and loss of confidence in nuclear power. The accident has had a good influence on the industry, because of the changes called for as a result. Perhaps one needs an accident to sharpen the mind. Engineering is a topic that progresses by learning the lessons of past problems; for example, the boiler code evolved from experiences with boiler explosions. In the field of human reliability, one has the idea that you can expect humans to be more reliable than 1.0 E-03. In fact, if you do not take the correct measures, even reaching this figure may be hard to achieve.

In this accident, the three main features are the failure of the PORV to close on demand, the design failure of not indicating that this is what could happen, and the failure of the operators, irrespective of the instrumentation failure, to diagnose that the accident was a small break loss of coolant accident (SBLOCA). The first two failures are design and construction faults. The last failure can be attributed to the failure of the training process, including the unavailability of good EOPs. In this latter case, the operators lacked the ability to understand the dynamics of PWRs.

The nonavailability of the auxiliary feed water only made the transient faster, but one could not argue that more time would have helped the crew. It was only the interruption of someone with a better understanding of plant dynamics that finally broke the cognitive lock-up of the operating crew.

Looking at the preaccident evaluation of human error, one realizes that there is a need for the HRA analyst to be more perspicacious than the operating crew and plant management. HDT analysis was done from an early viewpoint and a later viewpoint. The HEP in the first case is estimated to be 1.0E-03 and the latter case it is 1.47E-01. In fact, the latter HEP could be even higher depending on the evaluation of the experience factor. The arrival of the person from TMI Unit 1 in the control room suggests that the average crew may be better prepared than the crew at the time of accident.

The difference in the assessments raises some concerns about the awareness of HRA analysts to perform meaningful analyses. If the analysts are naïve, it is possible that they will miss certain issues and proceed to perform a "standard" analysis. Another thing to point out is that HRA methods predictions are based on average crews. One might want to understand if there is a large variation in the quality of the crews and operators and could this negate the findings?

9.4 CHERNOBYL NUCLEAR ACCIDENT

As stated in the accident description in Chapter 8, this accident is not really due to an operator error but rather is due to the combination of management decisions at a very high level to try to extract energy from the plant during shutdown and the failure to understand the dynamics of this type of reactor under these conditions. The operators were placed in a position of running an experiment without knowledge of what to expect. Additionally, there was a delay in the timing of the experiment because of a grid condition and the need not to take the plant offline. This, in turn, compounded the problems, because the only experienced operator who might have helped was replaced by an inexperienced operator. It is doubted that even the experienced operator could have prevented the accident. So clearly, given the circumstances, once the experiment was agreed on, the probability that an accident would occur was close to 1.0.

The effect of the delay not only affected the availability of the "experienced" operator but also affected the Xenon concentration, which made things worse. If one analyzed the probability of an accident ahead of time, ignoring the dynamics of the situation, one would predict a low probability. However, once it is realized that the organization had no idea of the stability of the core under these conditions, the chances of an accident jumped.

9.4.1 ANALYSIS OF THE ACCIDENT

The quantification of the crew probability of failure, assuming that the problem was limited to crew before the experiment was to be carried out, could be assumed to be about 1.0E-02 to 1.0E-03, in other words, quite low. This is built upon the assumption that the experiment was well understood and the procedures for carrying out the work were well developed and tested. Clearly, this was not the case, and the probability of executing the test without it leading to an accident was low. It appears that the organizing group had no idea of the risk and had no idea that the experiment could lead to the worst NPP accident in the world. Carrying out a risk benefit analysis without understanding realistically what the risk is, is meaningless. In this case, the before HRA would yield the wrong answer, because it was built upon inadequate information. One has to ask the question of the PRA analysts, do you really understand the situation before trying to quantify the process?

The answers obtained from the HRA analysis are that there are two prospective answers, 1.0E-02 to 1.0E-03 and 1.0. The first is the HRA contribution that the normal plant situation holds, and the second is under the assumption that the organizers of the test had no idea of the consequences of their decision. The first is that it is possible for the crew to make an error that leads to an accident, and the second is that the accident was inevitable. The prudent step in this would be to assume that there might be an accident and then look at the risk benefit. Under these circumstances, it is likely that the test would not have been attempted, even if the damage assumed to have occurred was slight.

9.4.2 SUMMARY

It appears from the analysis above that if one conducted an analysis assuming that the experiment was well conceived, the probability that the operators would cause an accident is from 1.0E-02 to 1.0E-03 and dependent on the skills and knowledge of the operators. The actual operator carrying out the operation was not very skilled, as far as this situation is concerned, so a higher probability of failure could be expected. However, this is not the real story. The core dynamics were not well understood, and the delay in carrying out the experiment only made things worse. Under these conditions, this could be considered a complete failure of management and the probability of an accident was high, even 1.0.

9.5 NASA CHALLENGER ACCIDENT

The space industry has not achieved the state of being a very safe business; in fact, it is just the opposite. Space explorations are inherent unsafe, because the normal steps taken to make a process safe are at variance with the needs to get a payload into space. The designers of space vehicles need to reduce weight, even more so than aircraft designers. The bigger the payload, the bigger the launch rocket has to be. The designer has to try to get reliability into the process by the choice of equipment rather than by redundancy and diversity, because each additional component weighs something. The exercise for the designer is to reduce weight and increase payload, and do so reliably.

NASA's record is either good or bad depending on one's view. It has succeeded in building several space vehicles and exploring space, including Mars. In this section, the focus is on the man-in-space program. In the case of the robotic space vehicles, it is not about losing persons, just the loss of equipment and time (i.e., loss of money). Every country that has been involved with rockets has experienced losses from the German V2 to now. This includes the United States, France, and Russia. Other countries acquired technologies from these countries or copied them, and this has either led to more reliable machines or they have not published information about their failures. Prior to the Challenger accident, NASA and the United States had space vehicle failures of one kind or another, such as those due to oxygen fires and rockets blowing up on launch.

The Challenger accident occurred because of the failure of a joint design involving O ring flexibility under cold conditions. Much has been written about this accident in the press and in books. There was a Presidential Commission (Rogers' Commission, 1986), and Feynman (1989) wrote a book on the topic. In this text, Chapter 8 covers the accident in some detail.

The accident progression is interesting, but here the prediction of the probability of the accident is what is of interest. The decision to launch and when to do so are the subjects of the prediction. To complete the story, the launch occurred and shortly afterward the seals failed due to the movement of the booster rockets under the force of their propulsion jets. Flames from the joint impinged on the oxygen-hydrogen fuel tank causing it to explode. The shuttle is made up of the combination of Orbiter, large fuel tank, and booster rockets. Due to this explosion, the shuttle crashed shortly after launch, and the entire crew died.

9.5.1 ANALYSIS

NASA goes through a formal process of preparedness for launch before a decision is taken. The engineering staff made the recommendation not to launch until the temperature conditions improved. The conditions were very cold. The engineering staff had previous information on the performance of the O rings under cold conditions, such as the presence of burnt O rings and indications of bypass burns.

The launch was affected by the fact that there was a teacher on board, and the U.S. president was going to make a speech on the "Teacher in Space" program. NASA needs to present a good case for the man-in-space program and for funding, in general, so there is a political aspect to NASA objectives. This meant that there was a need, in the minds of the NASA administration, to show that they could carry out a launch on time. The shuttle program had reached a degree of maturity and robustness, so there was every need to launch. To try to counter the advice from the engineering staff (engineers are mostly conservative), NASA approached a vice president of the manufacturers of the solid booster rockets for advice on whether he thought it was acceptable to launch. Naturally, he supported the firm's main client in their decision.

So how can one quantify the decision process? One could say that the NASA management was taking a calculated risk, but we are about to find out whether this was reasonable. In retrospect, it was not, because the mission failed. There are three decision states: the engineers' opinion, the VP's opinion, and the launch management's opinion. The HRA method that could be used here is CREAM, because Hollnagel provides estimates of interpretive faulty diagnosis and decision error. The faulty diagnosis would have applied to the engineers had they made an error; the decision error applied to both launch management and the manufacturer's VP. NASA launch management ignored the engineers' decision and accepted that of the VP, because it was favorable to their decision. The key decision was that of the launch management given the data on the joint coupled with the cold conditions to launch. The probability that the joint would fail might be as high as 0.5 under these conditions. The probability of the decision to launch could be the sum of the NASA and VP's probabilities. The probability estimated from the use of CREAM estimates would yield a mean of 2.0E-01 plus 1.0E-02 = 2.1E-01. The upper bound is 0.7, and the lower bound is 9.1E-02. The need to match the president's speech would drive the value toward the upper limit.

9.5.2 Summary

It appears that the Challenger accident was inevitable and not really a calculated risk given the data on the O ring characteristics. The human error of the launch management team coupled with that of the manufacturer's VP was high. A figure of 0.5 or 0.7 is a coin toss, so this is not really an informed decision of what to do but just a hope that all will go well. It did not. Management decisions should not be a case of coin tossing.

It would seem that both NASA management and the VP of the manufacturer's organization were irresponsible, given the considered opinion of the engineers and the presence of proof of a basic fault with the design of the joint under these conditions. The review by the commission did not say that the management team was right, and this was just an unfortunate accident that one could not predict. Maybe the numbers above are wrong, but at least they indicate that more accounting should be taken of what the people closest to the problem believe.

9.6   NASA COLUMBIA ACCIDENT

About 18 years after the Challenger accident, the Columbia accident happened. It involved the same organization—NASA. People have asked the question, has NASA learned from the organizational deficiencies revealed by the President's Commission under William Rogers? The answer is probably not. The accident is covered in some detail in Chapter 8. The main features leading up to the accident are as follows: the large fuel tank that supplies fuel to the main rocket engines that power the Orbiter is coated with an insulation material to prevent the fluid (liquid oxygen and hydrogen) from warming up during the time between tank filling and shuttle launch; the insulation material can detach during the launch, and there are preferred locations from which the foam detaches; the insulation during the launch phase can hit the Orbiter and cause damage to various parts of the Orbiter; the Orbiter is covered in heat-resistant tiles, and if the insulation hits the Orbiter, the tiles may become detached or broken; and the leading edges of the wings are made of carbon sheeting that is fragile, and the insulation could cause these carbon units to break.

The history of shuttle operations indicates that pieces of insulation have been seen to break off and bounce on parts of the Orbiter. On return from space, the Orbiter goes through a thorough inspection, breakdown, and repair. Tile damage has been reported, indicating that the insulation is not without strength to damage heat-resistant materials. The carbon leading edges are particularly fragile, so that a hard object could damage them. After the Columbia launch, large pieces of insulation were seen to hit the Orbiter, but the visual recording means did not clearly show the damage to the wing of the Orbiter. It was later determined that there was a large hole in the left wing of the Orbiter.

Many discussions went on about what to do about this hole. It appeared that not much could be done, because NASA was completely unprepared for this accident. Many ideas were advanced, including how to repair the wing, how to inspect it, and whether or not it could make it to the ISS. It was even suggested that a second Orbiter be launched to approach Columbia and carry out a rescue. Because Columbia was the oldest Orbiter, it did not have docking capability. The final result was that Columbia returned to earth and broke up and all the crew members died. This is another sad story of another failure by NASA, a failure to consider potential accidents and be prepared for the worst. Many of the ideas advanced during the analysis phase following the accident have been acted upon after the fact. The question is, why were these not considered before?

9.6.1 ANALYSIS

This appears to be another case of not taking the correct decisions prior to launch. Given that one was going to launch, what were the risks involved in this? Clearly, there was a risk connected to the insulation and its friability. Past records indicated that the insulation detached during launch. Somehow it was believed that the weight of the detached sections was low and the possibility that even the fragile carbon leading edges were strong enough to resist damage.

Interestingly, NASA subsequently carried out tests projecting hunks of insulation at the carbon units and they failed. Was this a great surprise to everyone? It appeared that if the insulation was coated with ice, its weight would be sufficient to rupture the carbon structure. Hence, a few early experiments could have been carried out and would have indicated a critical problem. One would need to be aware of the occasions when the icing conditions were such that there was a need to prevent the formation of icy foam prior to launch. It is not proposed to go through all of the other changes and modifications stemming from this accident.

This appears to be a failure by NASA to consider information related to shuttle operations and at least consider the consequences of such incidents. It could be that the organization was lulled by the idea that if it has not gone wrong before, then it will not go wrong in the future. Perhaps the conditions for the accident had just not occurred in this form before. The first inspection might be to consider what conditions lead to the formation of heavy foam pieces that might detach from the large fuel tank. It appears that pieces of foam had become detached in the past, so it was just a question of when one of the right size and condition would become detached.

The probability of the NASA management making the right or wrong decision depends on the information available to them to make a decision. One could argue that they did not have any information on this type of failure, and their decision to go ahead was correct. Another thought is that they were in this position before and all had gone well, so there was no difference this time than last time; therefore, they were correct to go forward. Equally, you could argue that they were lucky last time that things did go well. What we have is a decision process in which there is an unknown actor present that influences the probability of success and failure. NASA management was unaware of the characteristics of this actor and appeared to do nothing to determine its characteristics. They were aware of the foam detaching from the tank but never seemed to conclude that more protective features should be introduced in case the damage became severe. Their failure was not to do a risk assessment to examine the probability of the foam being great enough to cause damage and then look at what one could do to ensure that the crew was returned safely to earth. Some of these have now been done.

The decision making here is similar to that of the Russians carrying out the experiments at Chernobyl. The management was making decisions about launches in the absence of real information about the risk. What one needed was a curve showing the probability of releasing foam blocks of various sizes and depicting the size of foam block that could cause carbon leading edge failure. It would have been nice to have a set of curves for various temperature conditions and possibly time exposure conditions.

This might be a case for expert elucidation to get estimates of the probability of large or icy foam blocks being released. One could see the advantage of launching on a different day, taking measures to prevent the release of blocks changing the insulation design.

It would be better to prevent an accident from occurring rather than going into number rescue modes, especially if they are questionable. The idea of a repair kit sounds good and probably effective. The astronauts have carried out a number of space walks and performed complex tasks on the ISS, on the Orbiter, and on the Hubble telescope.

The decision making in this situation appears to be another case of coin tossing, as far as the block detachment phenomenon is concerned. Without data, one is assuming that the blocks will not release or is rather hoping that they will not. If the probability of blocks of significant size being released was high, then this accident may have occurred earlier. It appears that it is not high. Small blocks have been released a number of times and had little effect.

9.6.2 SUMMARY

The accident indicates the difficulty of making decisions in the absence of data defining the probability of an accident or at least the cause of an accident. One cannot carry out a risk–benefit analysis without knowing the probability of an accident. Here NASA did not appear to believe that large icy blocks would be released at any time, so there was no downside to launching the shuttle. It turned out that there was a risk that large icy blocks of insulation could be released. As far as one can understand, no steps were taken to see if block size could become larger and cause damage. Until the tests carried out after the Columbia accident, it was not clear what size of block and what kinds of characteristics could cause the failure of the carbon leading edge. Of course, there may have been tests in the past with nonsignificant missiles. In practice, the decision making in the absence of data was a coin-tossing situation that had worked out well in the past.

9.7 TENERIFE AIR TRANSPORT ACCIDENT

This accident is interesting in that the planes were scrambled to a local airfield unused to accommodating large planes (747s) following a bomb explosion at the Las Palmas airport. Las Palmas is the airport the planes would typically use. Under these circumstances, the tower personnel had to maneuver the planes so that they could eventually take off and resume their scheduled flights. Clearly, this was a rare situation, and the ground staff was quite unprepared. This means that the possibility of error was high. The details of the accident are given in Chapter 8.

The accident starts with one plane (Pan Am) missing a taxi off ramp and proceeding down the main runway to the next off-ramp. The other plane (KLM) was about to take off. A discussion occurred between the tower and the KLM pilot, during which the KLM pilot thought he was given clearance to take off. The confusion occurred between the tower and the pilot because of language differences (communication difficulty), and the consequence was that the KLM pilot took off and later the two planes collided. In addition, it was reported that there was voice distortion associated with the 747 radios. Visibility was not good, so the KLM pilot did not see the Pan Am plane until it was too late.

9.7.1 ANALYSIS OF THE ACCIDENT

The influences at play here are organizational deficiencies (tower personnel unprepared for this situation), Pan Am pilot failing to take the correct off-ramp, and KLM pilot having communication difficulties with the tower. The organizational difficulties are the prime items in this case, because the communications problems should have been understood by both pilots, and care should have been taken to go over instructions; the same should be said for the tower controller.

The key organizational issue was the failure to realize that this was an unusual situation and that extra care was needed to check that actions directed to be taken were, in fact, monitored and controlled, not to just assume all was going ahead safely.

There is enough blame to go around. Unfortunately, the plane crews and passengers were killed or injured. If one was going to attempt to predict the accident probability ahead of time, the initiating event was the bomb going off in Las Palmas. This is a low-probability event, and this is part of the problem. If it were a high-probability event, the authorities would have planned for a diversion of the planes to Tenerife and thought through the whole exercise. There was little to no planning on how to accommodate all of the planes and how to deal with the movement of the planes leaving the airport. All devolved on the local tower personnel, who did their best, but it was not good enough. The initial conditions for the accident were the KLM flight waiting at the end of the runway and the Pan Am plane moving down the runway preparatory to exiting on the access runway. The first error (HE1) was made by the Pan Am pilot who missed a sign and therefore continued down the main runway. The second error (HE2) was the failure of the tower personnel to notice that the Pan Am flight had proceeded beyond the correct ramp. The third error (HE3) was the miscommunications between the KLM pilot and the tower, which led to the KLM pilot taking off down the main runway. The combination of these errors led to the accident (see Figure 9.1).

The human error (HE1) attributed to the Pan Am pilot was that where he fails to observe the correct runway exit sign. It is surprising that he would do so because airline pilots are normally aware of runway signs. It is not known if the signs at Tenerife are any different. The second error (HE2) is that made by tower personnel in not controlling the Pan Am's progression and being alert to its noncompliance with directions. The last error (HE3) is due to communication difficulties between the Spanish tower and the Dutch pilot of the KLM 747.

If any of the steps was successful and some action was taken, then the chance of an accident would have been heavily reduced. Clearly, HE1 is an observational deficiency. It is not likely that the Pan Am pilot was not paying attention. There is no hard decision to make. It could be that signs were of a size to be compatible with small feeder line airplanes, and this could have caused a problem, but there was

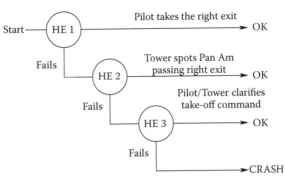

FIGURE 9.1 Human errors: Tenerife accident.

no mention of this in the reports. One must conclude that the pilot missed the turn because of observation deficiency.

For this study, it was decided to use the extended CREAM (Hollnagel, 1999) method. In CREAM, the wrong identification HEP is given as follows: lower bound 2E-02, basic value 7.0E-02 (mean), and upper bound 1.7E-01. The second error was another observational error, a failure to make the observation, and CREAM gives the same numbers as above. The last error was a failure in understanding the content of what was being said. CREAM does not cover this directly, but you could say that it was a failure to interpret the statements correctly and the nearest is delayed interpretation, the HEP values are lower bound 1.0E-3, basic value 1.0E-02, and upper value 1.0E-01.

The values could be modified by shaping factors of one kind or another. It would seem that the pilot of the Pan Am flight could be very relaxed at the time of missing the sign, and after missing the sign, he would become more concerned, but the error was made by that time. So the HEP1 value should be left as picked. The situation of the second HEP is again about the same as the first, the tower person is unaware that there is a problem; therefore, there is no shaping function contribution. The last case is HE3, and both parties would likely be stressed by the lack of understanding, but it is not very reasonable that this would affect the pilot's decision. This is a decision error based on poor communications. Using CREAM, it would seem that "faulty decision" under interpretation errors is the best fit. The corresponding numbers are 9.0E-02, 2.0E-01, and 6.0E-01. If we combine these errors, the result is, considering the basic values, equal to $7E-02 \times 1.0E-2 \times 2.0E-01 = 1.4E-4$. If one considers the upper bounds, the result is 1.0E-02. However, if one does not take account of the monitoring function of the tower, then the numbers are 1.02E-1.

9.7.2 SUMMARY

The prospective analysis of this accident indicates a severe underestimation of the potential consequences of failing to plan for this situation or even the Spanish authorities considering the impact of the scramble of planes to Tenerife. Sending a number of 747s to an airport used to handle a small number of feeder planes is going to cause some chaos at Tenerife. All of these difficulties could have been overcome if some among a number of things that went wrong had succeeded.

The failures included the failure of the Pan Am pilot to take the right exit, the failure of the tower to monitor the Pan Am's deviation, and the failure of the KLM pilot to be cautious in the light of communication difficulties. It is believed that the KLM pilot had no idea that there might even have been another plane on the runway. The analysis above indicates that there should have been a number of barriers (however, not physical) to prevent the accident, even in the chaotic situation. The really key person or persons to prevent the accident were the tower personnel. They were probably relatively inexperienced and did not supervise the progress of the Pan Am 747 to ensure it was where it should be at all times. Someone should been directed to perform this duty, either in the tower or on the ground locations near the exit ramp. One can see how the estimated numbers increase if the monitoring functions are omitted, from 1.4E-04 to 1.02E-1.

9.8 JFK AIRPORT ACCIDENT

This is an aircraft accident involving two planes: one an Airbus 300 and the other a Boeing 747. This accident arose because of the wake interference between the Boeing 747 and the Airbus 300. The Boeing is a big, heavy plane, and the vortices given off by any plane are related to the weight of the plane. The bigger plane has a bigger wake. The wake from any plane can affect planes following it in its airstream. The weight of the 747 is 5.5 times that of the Airbus 300. According to the NTSB, the accident was caused by human error, and it was that of the Airbus pilot.

The Airbus was affected by the wake from the Boeing, however according to the NTSB, the pilot should have been able to cope with the wake effect. The Airbus pilot's reaction to the wake led to the failure of the fin and rudder assembly of the Airbus, and the plane crashed killing all the crew and passengers plus five persons on the ground. Of course, the role of the NTSB is to try to find out the cause of the accident, and the pilot took the actions; no one else was directly involved. If you looked more closely, you could ask why the pilot would do this. The pilot is unlikely to choose to die along with his crew and passengers.

In order to understand the situation more clearly, if that is possible, one must see things from the pilot's viewpoint (see Dekker, 2005). This is not to say that pilots do not make mistakes, but sometimes the context of the situation is the contributory cause of the accident. In these cases, the pilot cannot defend himself and explain how this situation came about. Clearly, the interaction between the Boeing's wake and the Airbus is a contributory cause, because if this had not happened, then the Airbus would have flown off without any problems. The other issue is the design of the fin and rudder assembly—a question was raised about its strength and the design principles used to design it. Last, there are questions raised about the American Airlines pilot training for such instances and the pilot's past experience as a military pilot. All of these things should be taken into account in order to do a prospective analysis of the event.

9.8.1 Analysis of the Accident

The initiating event is the Airbus being affected by the Boeing's wake. The pilot's reaction is the human event, the failure of the fin–rudder assembly is the mechanical failure, and the consequence is the crash of the Airbus and the deaths of the people. The probabilities to evaluate are the initiating event, the human error, and the equipment failure due to maneuvers of the pilot.

The initiating frequency could be assessed by taking some number of flights per year from Kennedy and what percentage could involve a heavy transport with a lighter plane taking off at close to the same time. Let us say there are four such events per year. The author does not know how many similar events really take place per year, because a few minutes may make a difference, depending on how long the vortices last and have a strong enough effect.

The information about the pilots' responses to this kind of situation varies—some are trained to take little action and others strong action. The first group allows the passengers to be inconvenienced and rely on the weathercock stability of the plane.

The second group takes action to mitigate the effects of the passengers and tries to control the plane during this disturbance.

The actions of the pilot are influenced by his training, as to whether or not he fits into the former or latter groups. It seems as though this pilot fit into the latter group and took strong actions. Really the only question is, did he take too strong an action, perhaps? So let us say that 30% of pilots trained this way would take strong actions and that 25% of pilots are trained to take strong actions. Then the probability for this event is $0.3 \times 0.25 = 0.075$.

The next calculation involves the failure of the fin–rudder assembly. Given the maneuvers of the pilot, the failure is likely to be high; let us say the probability this fin–rudder assembly will fail is 0.1. There will be a failure under these circumstances of 1 in 10. Of course, dynamic strength tests could help determine this assumption. The other variable to consider is the reduction in vortex effect as a function of time, because the take-off time differences can affect the dwell time of the strong vortices and their influence on the following planes.

A major item for economical aircraft performance is weight. A designer is always looking to reduce the weight of a plane, and hence the move toward the use of fiberglass or other high-strength, low-weight materials. There has to be a balance between weight and strength, it has been suggested that the fin–rudder assembly of the Airbus was on the light side. Therefore, the use of a maneuver appropriate for an earlier designed plane may not be suitable for the Airbus 300.

The overall probability of a crash is given by the number of possible events per year times the probability the pilot will be in the group that will use the strong response action times failure of fin–rudder assembly given the strong motion. The numbers are $3 \times 0.075 \times 0.1 = 2\text{E-}02$/year or 1 every 50 years at Kennedy. So if there are 50 airports around the world with a Kennedy-like aircraft environment, there could be one such accident per year.

9.8.2 SUMMARY

An outline of the Kennedy aircraft accident is given. The analysis indicates the possibility of an accident like this at Kennedy every 50 years given the assumptions made during this analysis. There are several contributory effects besides the pilot training and his strong reaction to the 747 aircraft-induced vortex. The timing of the Boeing 747 takeoff relative to the smaller Airbus, and the "weakness" of the fin–rudder assembly design were contributory causes.

9.9 BHOPAL CHEMICAL PLANT ACCIDENT

One of the world's worst accidents, if not its worst, took place in Bhopal, India, at a pesticide chemical plant. Many thousands of Indians died in a painful manner, and many were permanently affected. It is asserted that the accident was caused by a disaffected employee, but other causes were advanced, including water passing via an opening on a header into a tank holding methyl isocyanate (MIC). The most credible cause seems to be the act of the disaffected employee. He is thought to have introduced water directly into an MIC tank, which then reacted with the water exothermically and released the chemicals into the air via a pressure safety relief valve. More details can be found in Chapter 8.

9.9.1 ANALYSIS OF THE ACCIDENT

If one assumes that the accident was the result of sabotage by an employee, can one predict such an occurrence? At least one can try to see if the circumstances are predictable. It has been said that the plant was due to be phased out seeing that the market for pesticides was not good. Union Carbide was considering plans to either close the plant or ship the plant to another country. Output from the plant had dropped, and people were concerned about being laid off.

This brings one to the estimation of probability of an accident in a prospective view. Given the situation as far as the mental health of the company was concerned, no safeguards were taken to prevent sabotage. Equally, the design of the plant seemed not to have an effective suppression system. Training staff to deal with such an accident did not seem adequate, but the staff tried hard to suppress the release.

If one was going to look at this situation credibly, one would review the history of industrial sabotage for situations similar to this. It is assumed here that the probability of sabotage was 1.0E-03/person/year at the time. For Bhopal, not all persons have access to key locations or knowledge to effect damage. The number of such persons is assumed to be 200. Therefore, the probability that sabotage is going to take place is 1.0E-03 × 200; therefore, the figure is 0.2/year as long as the plant is in this state. For a 2-year period (assumed), the probability is estimated to be 0.4 or the probability of a coin toss.

Now one can perform a risk–benefit study to see if the cost of putting in safeguards is worthwhile. It would appear to be obvious that if you kill 200,000 people plus injure others, then the benefit is very likely to outweigh the cost. The cost would cover enhanced security coverage of certain parts of the plant and an improved chemical suppression system to prevent releases from the plant affecting the public. Of course, Union Carbide might consider the cost to be too high given the circumstances and then choose to shut the plant down immediately, but at least this would be better than running the risk of massive deaths of Indian people.

9.9.2 SUMMARY

Analysis of the accident at Bhopal has been done in a prospective manner. With the assumed process and numbers, which could have been done by Union Carbide before the accident took place, it is clear that it is possible to inform management of the risks involved in running such a plant under these conditions. The potential threat of sabotage should have been considered as real. From the information seen in reports of the accident, the possibility of plant closure has already been seen to cause some problems with staff. If you are running a plant producing something like pump components, one might choose to ignore the threat of sabotage. However, when dealing with something as dangerous as MIC, the management should be very cautious and install preventative measures or decide to shut the plant down. The probability of this accident was predicted to be very high; therefore, the risk is high that a large number of people would be killed and injured, and that was the case.

9.10   TEXAS CITY OIL REFINERY ACCIDENT

The Texas City Oil Refinery is another chemical-related plant that seemed to be poorly run; see the more detailed report in Chapter 8. There were reports of fires before this fire. It also had instrumentation and maintenance issues. The accident was the result of a progressive series of errors and failures. The start of the accident occurred when a piece of equipment was being brought into service after maintenance. The equipment was a raffinate splitter. Oil is introduced into the raffinate unit and it is reformed and the product is sent to other parts of the refinery. The accident started with the crew overfilling the unit.

Depending on the reports, either the high-level alarm failed or the crew failed to act on the alarm. The crew continued to fill the unit and it overfilled. The crew responding to the release from the unit shut off the feed and opened the line to a blow-down tank. The amount of fluid passed to the blow-down tank exceeded its capacity, and fluid leaked out and down the tank and flooded the nearby area. The fluid is highly inflammable, and it ignited and led to explosions and fires. People were killed and injured. One issue was that caravans for service staff were in what might be called the exclusion zone. Persons in these caravans were among those killed and injured.

9.10.1   Analysis of the Accident

This is an interesting case to analyze prospectively. There are many things to think about and how to account for them. Often at the beginning of a PRA project, a number of tasks are defined, such as to analyze the state of the plant, to look at operating data, to look at maintenance of equipment data (failure rates and replacement of parts, etc.), to understand training programs and procedures, and to look at the quality of management (by interviews). In this case, if one had carried out such a review, it would have been seen that maintenance was poor, instruments seemed not to be working, the design philosophy of the plant was not integrated, and the top management were in London making demands for cost reduction at a time when money was needed for both operational and safety reasons. Questions would also be asked about the nearness of unprotected personnel to the plant that used volatile fluids that burned or could even explode. The previous statements were derived from the accident reports.

An accident event sequence was developed during the description of the accident, and it shows the main features of the accident. A similar diagram would have been developed as part of the PRA study. The initiating event was the filling of the raffinate splitter unit, and one would have thought the operations crew would have monitored the filling process. Once the level reached the alarm limit detected by a level sensor on the unit, an alarm should have sounded, and the crew should have acknowledged the alarm, canceled the alarm, and then taken action. This did not happen. The crew should have been watching the level rise on an indicator and seen when it reached the limit, even if the alarm did not function. If the crew was not monitoring the level indicator, they should have done this from the beginning of the fill to see all was proceeding as planned. If the level did not move at the beginning, the filling operation should have been stopped, but it was not. Therefore, the crew was not paying attention at the beginning or the indicator was not working. If the alarm did not work, the crew still should have known how long the fill operation should take. If it took much longer, they should have stopped the fill, but they only intervened once the raffinate column released fluid via a relief valve. The crew then stopped feeding the column and dumped the liquid to the blow-down tank. It appears that the amount of fluid in the raffinate column exceeded the capacity of the blow-down tank, and it, too, released volatile oil products into the environment. These petroleum products caught fire and exploded.

There appear to be issues concerning the following: equipment maintenance, personnel training, operational checks of instrumentation, operating procedure deficiencies, and design deficiencies. Behind all of these deficiencies is management. People can make mistakes, so some part of the accident progression can be laid at the feet of the operating personnel, for example, the failure to check whether or not the level instrumentation was working. This is almost the first step in being an operator. Of course, it could be the people were never trained to perform this duty. The vigilance and awareness of the crews were lacking.

The set of failures might have included the following:

1. Alarm, level indicator failed
2. Operator failed to monitor level
3. Operator failed to hear alarm
4. Operator failed to stop feeding the column
5. Operator failed to notice fluid overflow from the column
6. Operator failed to dump fluid to dump tank
7. Operator failed to notice dump tank overflowed
8. Management failed by housing staff inside the exclusion zone
9. Management failed to ensure maintenance of critical equipment
10. Management failed to select and train vigilant crews
11. Management failed to install useful procedures
12. Design of a dump tank too small for duty
13. A flare-off system was not provided on the dump tank (It might not have been useful anyway, because it is likely to be undersized as the dump tank was undersized.)

The operator could make mistakes by not monitoring the column but should have been warned by the level alarm. If the operator failed to take action then, the column overflow would have warned them, and it did. If they then acted to dump the entire fluid contents of the column into a correctly sized dump tank, there would have been no overflow. The effect would be that the nearby personnel would not have been killed or injured, provided the dump tank had been designed to accept the entire contents. There could have been a number of barriers (alarms, operator actions, and tank capacity) to prevent what happened.

It is possible for the operators to make a mistake and miss an alarm. If the operators' error for this miss was estimated to be a value of 7.0E-02, the HEP for isolating the feed and opening the valve to send the fluid to the dump tank following the indication of a release from the column has been estimated to be 1.0E-02. The overall error for failing to isolate the feed and open the dump line to send fluid from the column to the dump tank could be estimated to be equal to 7.0E-02 × 1.0E-02 = 7.0E-04. This is even missing the high-level alarm. The second action is considered to be independent of the first action.

These estimates were derived from CREAM (Hollnagel, 1999). If the error was due only to operator error, approximately 1 in 1000 operations of this kind would have to take place. Given that the instrumentation worked and the plant emergency dump tank was more conservatively designed, the accident would not have occurred. It could be that the human errors are misestimated; for example, the HEP of the operator missing the alarm might be higher than 7.0E-02, but it is not likely that it would be much higher than 0.1, which is about the same as given. The operator's actions in isolating the feed and opening the dump line were carried out under pressure during the accident. The estimated HEP value of 1.0E-02 seems quite reasonable.

9.10.2 Summary

Of course, not all of the above errors took place, because it is not clear if the equipment or the operators failed. The operators did stop the feed to the raffinate column, but only after it overflowed the column. They did notice that fluid overflowed both the column and tank. There were things that the operators could not know, like foreseeing that the dump tank was too small. If they had known, they might have dumped only some of the fluid and not continued to dump. Here the procedures were not spoken about in reports, which suggests that they did not know that the dump tank was undersized for this accident. The dump tank was probably not designed to accept the complete fluid contents of the column. It is hard to tell what error is dominant. The management probably is the most culpable, and the operators may have been undertrained and lacking experience. Again, it was the responsibility of the management to fund the training program. The management controlled key items like maintenance, test, operator training, and design. Maintenance governed the availability of the instruments. The training and procedures determined how the operators functioned. The management had a key role in the plant design. This accident is very much like the Chernobyl accident in that if the equipment functioned correctly, the accident would most likely not have occurred.

9.11 PIPER ALPHA OIL RIG ACCIDENT

The Piper Alpha Oil rig accident was a severe accident involving fires and explosions stemming from the release of oil and gas from not only the Piper Alpha rig but also other rigs that fed fluids into Piper Alpha. The British government requested an eminent judge (Lord Cullen) to hold an inquiry into the accident.

This accident pointed out many of the safety issues associated with oil and gas rig accidents. The issue is that there are a number of people working in a restricted space surrounded by the sea and exposed to the possibility of fires or explosions. The crew cannot just walk away; there have to be evacuation pathways, shelters, and escape routes to leave by boat. Piper Alpha was designed with oil fires in mind, but not so well for dealing with explosions. The protective aspect of rescuing the crew when the rig was on fire presented great difficulties. Also, providing shelter for the crew and then having the capability to leave the rig is difficult.

9.11.1 ANALYSIS OF THE ACCIDENT

Now discussion will focus on the aspects of the accident that affected accident initiation and progression from a human reliability point of view. The accident was a very simple accident in terms of its complexity. The consequences of the accident were quite bad because of some issues associated with the design of the rig and also the material being drilled and transported. This accident has much in common with the Texas City accident. It shows how concerned one should be when dealing with volatile and explosive fluids, like oil and gas.

The details of the accident are covered in Chapter 8. The Piper Alpha rig was the central node in a group of rigs transporting oil and gas back to the mainland in the United Kingdom. Gas was being pumped, and the pump being used failed. This presented a problem for the operators, because the management of the rig was concerned with continuing to pump gas. Stopping would lead to losses in revenue, so the pressure was on the operators to return to pumping. The system was a double-pump system, one main pump and one standby for circumstances such as this. Unfortunately, for the operators, the other pump had been withdrawn from service so that a pressure safety valve (PSV) could be calibrated. What the operators did not know was that where the valve was the section of pipe, a blanking cover was placed over the aperture. The blanking cover was not a pressure barrier, only a cover to prevent the ingress of contaminants.

The operating staff was not aware of these details that affected the operational capability of the pump. It appears that the records affecting the pump and the valve were held in different locations. The operational staff checked on the functionality of the pump and found that it was working the day before. There were no labels on the pump to prevent its operation, so the operations group decided to return it to service. All appeared to be all right at first, and then the blanking cover did not hold the pressure and gas escaped into a room. Later, the gas exploded and destroyed part of the top structure. A fire started and things got progressively worse. The fire suppression system that could have helped was shut down at the time because divers were working on maintenance operations in the water, and it would have been unsafe for them to turn on the fire suppression and flooding pumps, so the valves were locked as a safety precaution. For this study, one is interested in how this accident started and what could have been done to prevent or stop it. Given the situation on the rig as far as the fire was concerned, there was little that could be done once the fire caught hold. People on the rig were in survival mode, and even that was difficult. Even ships could not get close to help remove people. On top of that, other rigs continued to feed the fire rather than shutting off the oil and gas. It took awhile to get things under control from just this aspect.

The analysis of this situation is dominated by the actions taken by an operator in switching over from a nonworking pump to a pump that had been available but was withdrawn from service because the PSV had been removed. One could quantify this act by using CREAM by deciding that this was caused by the incorrect decision made in the absence of information about the state of the other pump and the PSV. The information available to the operator about the operability of the pump was available, but the information about the associated PSV was not located with the pump data, but in some other location. Now one is faced with what model of the situation is given by CREAM. It is a wrong decision, but the information available to the operator was incomplete. The nearest item in the CREAM tables is a decision due to incomplete diagnosis, and there are others. None of these are right, but let us examine the numbers: the mean value is 2.0E-01 and the upper and lower numbers are 6.0E-01 and 9.0E-02.

If one applied another HRA method, one might choose HDT. Applying HDT, one needs to consider the set of associated influence factor characteristics such as training, experience, procedures, MMI, and communications. In this case, one might argue that the training was not good, because the crew did not do a walk-down to see if the other system was aligned correctly or if there had been any maintenance tags on other equipment. The procedures were questionable, because they did not call for information on the availability of standby systems for the personnel in the control room. The MMI also was deficient in that there was no indication of the nonavailability of the standby system. Communications between the maintenance personnel and the control room crew were lacking. Many of the items fall into the same failure to have in place ways of operating the rig. There are some design issues as well, for the MMI was poor, because the control room ought to be the place for sufficient information on the status of equipment needed to operate the rig safely and effectively.

The HDT model for this situation would have the five IFs mentioned above. The weighting between the IFs might be different, but the accident report does not help to distinguish that, so equal weighting will be assumed (i.e., 0.2). All of the quality factors (QFs) seem to be poor. Perhaps this is a little severe. For this industry, maybe many of these things are acceptable, but for safety-related industries, most of the QFs would be judged to be poor. Given these ratings, the estimated HEP is 1.0, which is not surprising.

It appears that the management control over the rig operation and design was poor. The management should have more effective control over the processes. The operations crew was not aware of the state of the plant. The procedures did not call for checks before turning on equipment. The recording of maintenance operations seemed to be on a per-item basis rather than on the availability of complete systems. The control operators were responding to economics rather than safety considerations. Furthermore, the other rigs were not immediately told to shut down pumping toward Piper Alpha and to isolate the lines. Isolation of the lines from the other rigs may not be sufficient for gas, because the line was pressurized and the gas would continue to feed Piper Alpha, unless there was a shut-off system local to Piper Alpha. Many other aspects of rig operation and design were defective, such as escape routes for the rig personnel, control over the fire suppression system from different locations, means of safe docking of rescue ships so that they can collect the rig personnel, even if the rig is on fire, and there ought to be a safe refuge for the crew to wait out the fire, if they cannot be rescued quickly. Although the control room crew should not have taken the steps that they did. The whole operation was poorly conceived. Ultimately, the management must be considered responsible.

9.11.2 SUMMARY

The assessed human error probabilities using CREAM and the HDT method gave about the same result: The rig was poorly designed and operated and, if some incident would occur, like the one that did occur, severe accident would be the result. Of course, judgment has been made of the conditions based on what has been revealed by the accident report. Principally, the information revealed was about the lack of controls over the actions taken by the control room crew to continue pumping gas without doing a detailed evaluation of the system before switching over to the other pump.

The question that remains after performing such an analysis is, would one be able to carry out a detailed evaluation of the rig operation to reveal all of the shortcomings seen in the report? This is a fundamental issue to ask PRA groups engaged in performing PRA studies.

If the PRA study revealed many of these shortcomings, how would the management receive the information? Would they "shoot the messenger" or say thank you and do something about correcting the situation? The lesson from NASA would lead one to conclude that they would enlist another person or organization to get a more favorable opinion. Not many people want to go to the board of directors and say that this study just revealed a whole load of problems that are going to take a lot of money to solve. The other point might be, why did you not know about all of this? One can see why accidents occur when it should be possible to prevent them.

9.12   FLAUJAC ACCIDENT, FRENCH RAILWAY

In a way, the Flaujac accident is about what can happen in a single-line rail system, with bypasses. An accident involving the meeting of a heavy transport train and a more lightly designed passenger train will always end with the passenger train suffering heavy damage with death and injury to the passengers. Whatever the detailed design of the track signal system, if it relies upon humans without the additional security of a mechanical or electrical or computer system, trains will inevitably have an accident, it is just a case of when.

The design of NPPs was postulated originally on the premises that any single failure will not lead to an accident. The single-track train system is not based upon this premises, but rather the human controller or controllers will not fail. It appears that a number of HRA methods have reinforced the idea that the reliability of humans is bounded, so that there is a reliability value that humans cannot achieve despite all attempts to design such a system. It is the author's opinion that this number is about 1.0E-03. So the range for humans is 1.0 to 1.0E-03. Perhaps under certain conditions, this number may be reduced by an order of 10.

The Flaujac accident is discussed in more detail in Chapter 8. The signaling system attends to the idea of involving more than one person to act as a check. One controller tells another if the train leaves a given station, and the other controller is supposed to know the schedule for that train as to where it will stop, but he also knows of the schedule and location of the freight train. The freight trains are supposed to pass when the passenger train is off the main track. This sounds good; however, one of the controllers who is very experienced has been replaced by a less-experienced person. The trains do not have the same schedule every day. On days when the schedule varies, the train's sequence is different. The controller needs to adjust the timing and location of where the passenger train may stop. On the day of the accident, the controller thought that the trains were following another schedule rather than the one the drivers were following. When the controller heard that the passenger train was in one location, he thought it was in a different location. By the time he worked out that he had made a mistake, it was too late, and the two trains crashed.

9.12.1 ANALYSIS

One could try a couple of different HRA models to try to estimate the HEP for this situation. The two methods considered are the HDT method and the CREAM extended model. For the evaluation considering HDT, one needs to determine the influence factors (IFs) and consider the quality of each of the IFs as understood by the context under which the controller is making his decisions. The IFs selected are training, experience, MMI, and procedures. The first two are fairly self-evident; the man–machine interface is the train schedule depicted for different dates. The procedures are the rules governing what the controller is supposed to do and what his duties are. In this assessment, the importance of each of the IFs is judged to be the same, which is 0.25. The range of QFs is 1, 5, and 10. One is the highest quality, five is average, and ten is the lowest. Two cases are considered. The first case is where the QFs are 10, 10, 5, and 5. For the first case, the HEP is calculated to be 1.5E-01. For the second case, the QFs are 5, 10, 5, and 5. For the second case, the HEP is 6.0E-02.

For the CREAM evaluation, the selection is faulty diagnosis, and the HEP central value is 2.0E-01 with upper and lower values of 6.0E-01 and 9.0E-02. These sets of values are quite close, with CREAM yielding the highest HEP. But the probability of an accident if determined by these figures is quite high. If one takes the CREAM number of 2.0E-01 per event, it means that with this person there would be an accident every five opportunities, every time this configuration occurred. This is much too high to be acceptable by the public. However, if there was a mechanical or electrical system that prevented two trains being on the same track at the same time with even a low reliability of 1.0E-02, then the combination would yield an overall number of 2.0E-01 × 1.0E-02 = 2.0E-03. For the more experienced controller, this HEP would be better than that of the novitiate. If his numbers are 1.0, 1.0, 5.0, and 5.0, respectively, for training, experience, and procedures, then the HEP for this combination is 4.6E-03, and this is significantly better than his replacement. If his numbers are combined with those of the device, then the numbers become extremely small.

9.12.2 SUMMARY

The HEP figures tend to confirm the likelihood of an accident given the combination of factors associated with the controller. It is not surprising that when one considers that the controller is not very experienced, so the experience factor is poor and the training factor is also poor, the rest is average. Compared with the more experienced controller, the first two are good and the last two are average. The single-line track demands the best of the controllers, and if there is any falling off, one can expect an accident.

9.13   KING'S CROSS UNDERGROUND RAILWAY ACCIDENT

The King's Cross accident was one that was waiting to happen. The postaccident evaluation showed that many fires had occurred in the past caused by lighted matches falling into the space under the moving escalator treads or steps. The space under the treads was covered by a mixture of grease and fibrous fluff. It is surprising that a big fire had not started before.

Truly, cleanliness is next to godliness—if you do not remove the flammable materials, the chances are increased, when the public drops lighted matches. It is realized that this is a serious situation, but one can measure the value of good management by examining the workplace. This process holds for all kinds of workplaces, including nuclear plants. Fire is a great hazard and needs to be treated carefully, as one can see by the Piper Alpha accident and this one. Fires are often started in the following manner: a pump gearbox leaks oil and the maintenance personnel use rags and fibrous materials to mop up the oil, and then leave the debris in place, just waiting for a spark to set it alight.

In this case, a fire did start, and some white smoke was released due to the stair steps burning. The treads of the escalator were made of wood and caught fire. This was just the beginning, however, as the staff and firefighters thought that the fire was relatively benign. Because of the white smoke, it was treated somewhat casually. Urgent measures were not taken to evacuate all passengers in the underground complex. Subsequently, the fire developed and grew. Then an unknown phenomenon took place called the trench effect to increase the effectiveness of the fire. In addition, movement of the trains through the tunnels caused air pumping. This increased flow of air helped the combustion process along. Sometime during the fire, the conditions changed, and the smoke became black. This change affected the evacuation process and the net result was that a number of people died.

9.13.1   ANALYSIS

The probability of a fire was always high once a lighted match was able to start a fire and the materials permitted it to grow. It appears that over the last three decades, there have been 300 fires. The expectation of a severe fire seems to have been low. One might hypothesize, why did a severe fire occur now? It is suggested that a more volatile mixture existed of oil-impregnated paper in the mixture and that this caught fire. The number of lighted matches per year falling had gone down because there are fewer smokers, so the probability of an event like this should have been reduced.

The key event appears to be the decision not to quickly evacuate the station and prevent people from debarking at the various platforms. However, this is easier said than done. King's Cross is a multiline station with some six underground lines involved. King's Cross is also known as King's Cross–St. Pancras station and is the busiest underground station in London. Since the fire, it has been undergoing reconstruction to meet the recommendations of the Fennell public commission (Fennell, 1988).

Estimation of the human error for this accident can be accounted for by problems associated with decision making and fire evaluation. It is not clear who made the decision not to take the fire very seriously and order evacuations. King's Cross is a very complex station with many interconnections; therefore, a fire can cause a lot of problems as far as the safe evacuation of passengers and staff. The fire started on the Piccadilly line escalators, which is the lowest line.

The HRA model used for this is the CREAM extended model. The HEP is calculated from a faulty diagnosis of the problem and yields a mean figure of 2.0E-01, with the upper and lower values of 6.0 E-01 and 9.0E-02, respectively. These figures are high, and maybe there are some ameliorating conditions. There are two effects that could cause the decision maker not to take a strong action: 300 fires had been extinguished before and the smoke was white rather than black and oily. According to the CREAM process, the experience and knowledge could have reduced the estimates by 0.5. The HEP central value would become 1.0E-01, still high. So given the fire was not quickly extinguished, it was inevitable that people would die because of the fire characteristics, availability of escape routes, and failure to both extinguish and evacuate early.

9.13.2 Summary

As previously discussed, fires are difficult, need vigilance, and should not be underestimated. The King's Cross accident indicated that the fire was underestimated. Sufficient care was not taken to remove the causes of the fire. One step was in the right direction, which was to ban the lighting of cigarettes, but clearly it was difficult to enforce. Some aspects of the fire were unknown to the people involved in making decisions, such as that the effects of the fire could travel downward and the trench effect.

Clearly, the probability of a decision error was high, a mean value of 1.0E-01 with an upper range of 3.0E-1, allowing for a weighting factor of 0.5. This seems more like a coin toss, like some of NASA's decisions. There needed to be some thought given to this type of situation, especially for stations like King's Cross, ahead of time. It always seems to be a case of too late.

9.14 COMMENTS

A review of the application of HRA techniques has indicated two things: There are situations in which current HRA techniques are limited in predicting the real probability, and there situations that have not been identified by industry personnel as highly risky. In the case of the latter, the organization only becomes aware of the fact after an accident. This is a little too late. Analysts must bear some responsibility for this as well, because they can use the skills and knowledge ahead of time based upon reasonable estimates rather than full-blown studies. It is not an either/or situation. Most of the calculations carried out for this analysis have been restricted and somewhat limited because of the information available. However, often one can see that something needs to be fixed at the plant or factory, or even the control of planes on the ground or in the air.

Often the situation is dominated by decisions made just before an accident or much earlier. The management in these circumstances is making decisions without understanding the consequences of their decisions. The operator is the person carrying out the action, but his action is conditioned by the combination of earlier decisions taken by management. In the case of NASA, the decision makers are the management, and often the advice of more involved personnel is set aside because of what is often justified as the "big picture." This kind of attitude is not limited to NASA.

The analysis of many situations can help management better understand the risks involved in their industries. HRA techniques can be applied as implemented above, using methods like CREAM to find a problem along with PRA techniques to understand the relationship between human errors, equipment failures, and consequences. Scoping studies can be valuable: in fact, one of the tools used in PRAs is screening to ensure that HRA studies are as cost effective as possible. These screening studies are conceptually what is recommended but applied to the whole PRA.

The HRA analysts should be involved with trying to understand more about the functioning of an organization, so that they can see the interrelationships between management and operations and how this affects plant safety. Too often, the HRA personnel take a very narrow view of their involvement. It should not be limited to providing a number, as this may be the least significant part of their contribution.

10 An Application of an HRA Method (HDT) to a Real Problem

INTRODUCTION

Earlier, in Chapter 6, various HRA methods were discussed. Among them was the holistic decision tree (HDT) method. This method has been used in a number of applications, but one of the most interesting was its application to space applications. It was used to improve on a prior application of an HRA method in the International Space Station (ISS) PRA. The majority of HRA applications have been in the nuclear power plant (NPP) business for obvious reasons. The concept of using PRA started with the nuclear industry because of the perceived risk of a large number of deaths in the case of a large accident. It was determined that human error could be a significant component of that risk; therefore, attention was paid to trying to quantify the risk of human error. The development of HRA methods and techniques is discussed in Chapter 6. There is a convention in the HRA field to discriminate between the older and newer HRA methods. The early methods were grouped under the title of first-generation methods and the later methods under the title second-generation methods. THERP (Swain and Guttman) would have been designated a first-generation method.

The basis of some second-generation models is that the context in which humans find themselves during an accident scenario is an important contribution to the determination of human error probability (HEP). The HDT method is such a method. The HDT model uses a tree construct to show the relationship between context and error probability. The context is the environment in which the operators find themselves and includes the impact of the procedures, man–machine interface (MMI), training, and so forth, on the operators.

10.1 DESCRIPTION OF THE INTERNATIONAL SPACE STATION

The ISS is a large orbiting space station made up of a number of modules. These modules have been assembled in space to serve as work stations and crew quarters. About 16 nations are involved in the project. The main components have come from the United States, Russia, Europe, Japan, and Canada. More information on the history of the construction of the ISS in space, the assembly of units, and the involvement of countries and personnel in the project can be found at the NASA Web site (www.nasa.gov). Pictures and artist renditions of the various stages in the assembly process can be seen by visiting the Web site.

The ISS is very much like a production facility running experiments in orbit. Its power is produced by solar panels, and there is an internal power distribution system. The station needs to be air-conditioned and pressure controlled for comfort and life support of the crew. The astronauts or cosmonauts wear pressure suits only for operations outside of the station. The suits are used while performing maintenance and replacement duties. Ground control and communications are via satellites.

The safety of the crew is a primary consideration, and if required, the crew can leave the ISS by the use of a Soyuz three-man capsule. Crews can be replaced by either Soyuz capsules or by Orbiter or shuttle. The Orbiter is used for resupplying ISS or transporting additional ISS modules. The Orbiter is used to transport replacement parts as required. Currently, there are two resupplying vehicles: One is a modified Soyuz (called Progress) and the other is a recently designed and used European vehicle (Jules Verne). Both are automatic docking vehicles. The Soyuz is limited to transporting three cosmonauts, either to ISS or off ISS. The Orbiter can carry more people; typically it can carry a seven-person crew. It is much more flexible than the Soyuz, but the Soyuz is a very reliable vehicle and has seen a lot of service.

There are two means of stabilization of the station: gyros and impulse jets. This gives the ISS a unique means of stability control that is diverse. The gyros are U.S. designed, and the impulse jets are Russian designed. This diversity paid off when the gyros failed and the jets were used to stabilize the ISS. A problem with the jets is that they use fuel, and the fuel has to be supplied from earth during crew replacement and provisioning flights. Multiple computers are used to perform all of the tasks in space, such as switch over of power sources, orientation of solar panels for optimal performance, and control of the fluid systems.

There are two groups of persons associated with direct ISS operations. These are astronauts/cosmonauts and flight controllers. The primary function of the astronauts is to perform four sets of duties: respond to rapid accidents (i.e., accidents that need actions taken quickly after initiation), be in control of experiments, support the flight controllers in detecting anomalies in systems, and engage in maintenance activities.

The flight controllers operate from the ground and are responsible for the correct running of the ISS. This function has to be carried out at all times, and they have to be available at all times. This is accomplished by manning different shifts. The controller teams consist of a flight director, in charge of the operation, a communications person (Cap Com, a carryover from the days of Mercury, etc.), and individual controllers in charge of various systems.

The controllers are responsible for monitoring and controlling the ISS systems remotely. The astronauts do not control the various systems; their role is not like the earlier space vehicle personnel, who had a lot of different responsibilities. Individual controllers look after various systems by monitoring and taking actions to ensure they are performing correctly. Information gained from the various systems in the ISS and control actions taken occur via communications channels to satellites and, hence, to ISS computers. The control signals are sent from the computer to operate switches to open and close connections in various systems, for example, if an electrical transformer is not behaving correctly, it is switched off and another is turned on.

Another important function performed by the controllers is to make changes in the orbit of ISS to avoid space debris.

The health of ISS is monitored at all times by the controllers. When the ISS crew sleeps, the flight controller numbers are reduced, but the ISS is always being monitored. The main control center is in Houston, Texas, but there is a standby center in Moscow. The Moscow center has been called into operation a number of times—once when a hurricane hit Houston, and another time during the failure of all three on-board computers. The Moscow center performs a critical backup function.

There is some limit to the amount of information available to the controllers, so on occasion it is necessary to engage the astronauts in assisting the controllers to debug a system to find out why it is not working correctly. A simple issue might be a plugged line in the fluid system, part of the thermal conditioning system. The controllers may see that there is a problem but may not be sure of its actual location or what is causing it. In these cases, the astronauts become the eyes and hands of the controllers.

10.1.1 OPERATING CONDITIONS AND ROLES OF ASTRONAUTS AND CONTROLLERS

Some of the roles of the astronauts and controllers have been covered above. Here one has to consider their roles during various accidents. The roles can change depending on the accident. If an accident occurs very quickly, then the astronauts will have to deal with it. If the accident is slow in developing, then it can be tackled by the controllers.

As in the case of most systems in other fields, there are equipment alarms of various kinds—red (alarms), yellow (caution), and blue (alerts). The division is as follows: the astronaut crew deals with the reds, the astronauts and controllers work together on the cautions, and the controllers deal with the alerts. As mentioned above, the controllers may call upon the astronauts to help with diagnostics. Over and above dealing with accidents, there may be times in which the only thing that can be done is for the astronauts to evacuate the ISS. The Soyuz capsules are provided as rescue vehicles, so the crew would move to the Soyuz location and enter the Soyuz and leave. The other thing is that the crew must never be cut off from being able to leave the ISS by a Soyuz, if there is an accident. The crew must position themselves within the ISS to ensure they do not get cut off by the source of the accident being between the crew and the Soyuz. Figure 10.1 represents the relationship between crew and controller actions and the alarm state.

10.2 ISS PRA: BACKGROUND

In line with a number of organizations, NASA requested that a PRA be carried out to be used to define the areas of risk associated with the ISS operations. The PRA study was carried out. The PRA was then reviewed by an independent review team to ensure that the results and the methodology were acceptable. This is a very usual process in the PRA field. Among the number of review comments was a suggestion to improve the HRA to base it upon the knowledge and experience of NASA. NASA has a lot of experience in the man–space field, going from Mercury, Apollo, Orbiter/shuttle, and ISS operations. The PRA review team felt that the HRA could be improved by using this experience and knowledge. It should be pointed out that

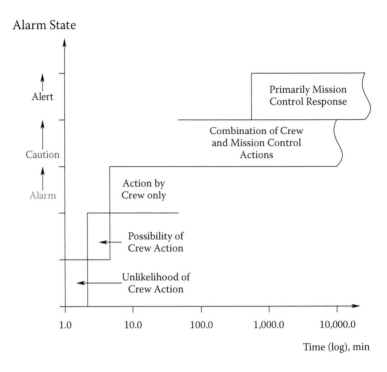

FIGURE 10.1   Action states of crews and controllers.

the original HRA was based on CREAM I (Hollnagel, 1998), which was considered to be too general to meet the requirements identified by the review team. A set of criteria was generated to be used in the selection of the HRA method. These criteria are as follows:

1. The approach used must reflect NASA knowledge and experience.
2. It should be a second-generation method. These methods incorporate the experience derived from the use of the earlier methods.
3. It should have been peer reviewed by a credible group of investigators.
4. It should have been applied by a number of different organizations.
5. It preferably should have been used in a number of different applications.

The answers to the above criteria related to the HDT method were as follows:

1. The HDT method is sufficiently flexible to be used for this application.
2. It is a second-generation method that has evolved from the observation of control-room crews responding to simulated accidents.
3. Among the various users, it was peer reviewed by the Hungarian Physics Society at the request of the Hungarian Atomic Energy Commission and found to be an acceptable method.
4. It has been applied by organizations in Hungary, France, Russia, and the United States.

5. It has been applied to different PSA Level One studies, including full power, low power, and shutdown modes and for internal and external events. It has also been used for a Level 2 study by the French to cover the use of "crises centers" during NPP accidents that could affect the general population.

After a review of possible HRA methods, it was recommended that the HDT method be used based upon answers given for the above set of criteria.

10.3   HDT METHOD APPLIED TO ISS

The important aspect of the HDT approach is that the human error rate depends on the context in which the operators function. In the case of the ISS, before proceeding, one needs to determine the context under which the personnel are operating during the various accidents. The context may vary depending on the characteristics of a specific accident. One has to determine the features of a specific accident and relate them to the specific influences that cause the HEP to reflect both the accident and its associated impact on the personnel. The context is made up of a number of influence factors, such as changes in displays, atmospheric pressure, and so forth, and the response was evoked from the personnel because of the situation by virtue of their expertise, experience, and training. The importance and effect of the influences depend on the situation. For example, the impact of the MMI may be different from one accident scenario to another; hence the resulting HEP will be different.

The HDT model is a representation of the following word equations relating the context in which persons find themselves, to the error probability:

HEP = Function (Context)
Context = Function (Influence Factors)
HEP = Function (Influence Factors, such as Quality of Procedures, Quality of the Man–Machine Interface, Quality of Procedures, Quality of Leadership, etc.)

The HDT method as applied to the ISS PRA, has been selected as an example to illustrate an HRA study. In this chapter, elements of the complete study are omitted, because they are specific to NASA. The general approach could be used for a nuclear industry or any other high-risk industry HRA study. There are some features that are specific to the manner in which the ISS is controlled. How responses to accidents are carried out by the astronaut crews and flight controllers are quite different than how they are carried out at a nuclear plant, although in the case of level 2 nuclear power accidents, the use of offsite personnel is somewhat similar.

From the beginning in the development of the HDT approach, a tree structure has been used to enable one to visualize the contextual relationships and how they affect the end-state probabilities. The contextual influence factors are depicted by the headings of a tree structure, like an event tree. The importance of any influence factor within the set of influence factors can vary from accident scenario to accident scenario. The quality of an influence factor can also vary. As shown in Figure 10.2, the main elements of the tree are the headings, the branch structure, importance weights of each heading, and the quality of each influence factor. The headings are the influence factors. The number under each heading is the importance weight of that influence factor for the accident scenario of interest. In one accident, the analysts judge the MMI as the most important effect and the other influence factors have a lesser effect. To explain the quality factor, one should look at situations in which the main influence is the MMI, but unfortunately the MMI design is not very good, and its quality is poor. If there are no compensating influences, then the resulting HEP is going to be high.

The HDT for a coolant loop leak is depicted in Figure 10.2. The numbers on each branch of the tree are the quality values (QVs) corresponding to the qualitative quality descriptors (QDs). The QDs rate the condition or "goodness" of each influence

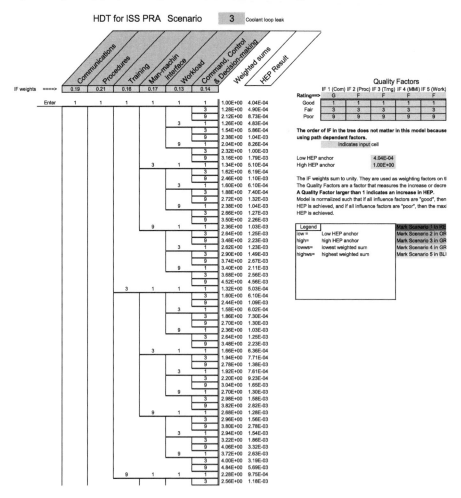

FIGURE 10.2 Typical holistic decision tree representation.

factor in supporting the actions of the ISS operations team. The quality values are the numerical increments in human error probability assigned to each qualitative QD. The QDs are shown on the right of Figure 10.2. The QVs shown below the QDs are also in yellow.

The tree starts like an event tree at a single point and then branches at the first heading into a number of pathways. The number of branching points equals the number of different quality ratings used for a given influence factor. In the study, three QDs are used for each influence factor. Alternative phrasing was used for these three descriptors, such as good, fair, poor; supportive, adequate, and adverse; efficient, adequate, adverse; good, adequate, adverse; or more than capacity, at capacity, or less than capacity. Therefore, each input branch splits into three output branches.

10.4 STEPS IN PERFORMING AN HDT ANALYSIS

The steps in performing an analysis are as follows:

1. Draw up a list of the potential IFs.
2. Sort the list into scenario-dependent and global influence factors. The HEPs are affected by either scenario (local effect) or global influence. The global influence factors will be found in every scenario. In the study, all the influence factors were global, and the same ones were used for all scenarios considered.
3. Rank influence factors in order of importance, and select the most important ones.
4. Select and carefully define the QDs to be used for each influence factor and draw the logic tree.
5. Estimate the importance weights of each influence factor. In this analysis, expert judgment was used along with the variability of judgments by the various participants as a measure of uncertainty.
6. Determine the upper and lower anchor values for HEP.
7. Using the QDs, rate each influence factor for each scenario.
8. Determine the incremental change in HEP associated with each QD rating. These are the QVs.
9. Calculate HEP distribution. The calculations are codified in a Microsoft Excel® spreadsheet submitted separately.

These steps are described in the following sections.

10.4.1 STEP 1: LIST OF POTENTIAL INFLUENCE FACTORS

In the ISS case, the various steps in the above process were taken in conjunction with NASA personnel so that the final HDT formulation reflects actual NASA operational experience. The knowledge elicitation sessions were conducted with NASA controllers, crew, and trainers to tap into NASA experience and knowledge. Equally, the same process could be used with nuclear power control room operators, instructors, and maintenance personnel, depending on the HRA requirements.

10.4.2   HRA: Plan and Process

The various steps laid out in the construction of the HDT model given above are mirrored in the HRA plan. A project plan should be drawn up similar to that done for the use of NASA expertise. The integration of plant expertise and HRA knowledge on the part of the HRA experts is an important element of this approach. Figure 10.3 shows the steps in the project plan for the development of an HDT model of the

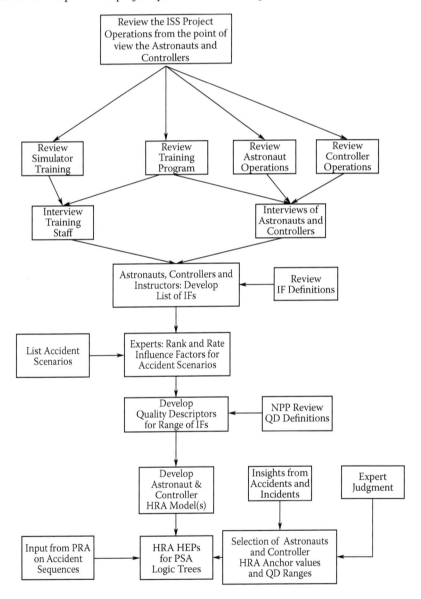

FIGURE 10.3 Overall project process to develop International Space Station human error probabilites.

control room crew responses to accident scenarios. The plan consists of a number of steps designed to inform the HRA experts about operations, use of procedures, operator training, and those plant design aspects that can affect human reliability in responding to scenarios.

It may not be possible to define a single HDT model to cover all of the accident situations that might arise. It is an accepted fact that the context varies according to the type of accident. In other fields, this can mean that the same variation in the quality of the IFs can occur, but essentially one is dealing with the same crews. The effect of this is that the HDT can remain the same, but the pathway through the tree structure can vary due to differences in each scenario and lead to different human error probabilities (HEPs).

In the case of the ISS, the importance of the IFs can change along with the quality of each of the corresponding descriptors. For example, for a specific accident, the crews may be more affected by training quality than by the man–machine interface quality. The crews in another accident may be more affected by the quality of the man–machine interface than by procedures, and so on. In principle, therefore, one might have a different HDT model for each specific accident scenario. It is not always practical to build such high resolution into an HRA. For the ISS HRA, five representative accident scenario categories were developed.

10.4.3 INDOCTRINATION PHASE

As part of the indoctrination process, visits were made to the NASA Johnson Space Center (JSC) to acquaint the HRA expert with the ISS operations. The early visits were undertaken to see the training simulator, astronaut training tank for training in extra vehicular operations, the internal arrangements of the ISS mockup, including the Soyuz capsule, and visits to the flight control room. The space in the Soyuz is very cramped with the three astronauts/cosmonauts sitting knee to knee. Time was spent witnessing various training sessions covering simulated accidents. A couple of things were noted about the simulator sessions, the sessions were quite long compared with power plant sessions. Power plant accident sessions last for half to one and a half hours compared to ISS sessions, which may last for several hours. The nuclear power plant instructors' involvement in the accident scenarios used for requalification training is quite limited, and they tend to play the part of auxiliary instructors. Equipment failures are included as part of the designed test scenario. For the ISS sessions, the main objective is often the training of the controllers as a team. The instructors design the scenarios and interact with the various systems to mimic the failures. Sometimes astronauts are included to cover the things expected of the ISS crew during an accident sequence.

As part of the HRA exercise, some time was spent trying to understand the ISS operations, and part of that time was spent at the simulator complex. The activities directed toward framing the HRA study take place with mixed groups of controllers, astronauts, and instructors in a room set aside for this purpose. Table 10.1 lists a summary of the activities featured during visits to the JSC facility.

Given that data were collected from each of the participants, a series of calculations is presented.

TABLE 10.1

Summary of Visit to Observe Scenarios and Cover Sessions for Determining Important Influence Factors

Witness multifailure scenarios.

Each day follow a scenario; gather the crews, controllers, and instructors together to carry out elucidation sessions to determine the main influence factors thought to influence crews and controllers when responding to accidents. Prior to each elucidation period, there was a short training program on what are influence functions. The participants are requested to list in their opinion what are the top IFs and a brief description of each.

Collect data sheets from participants.

The data were assembled and presented to each group.

Presentations on actual incidents.

Following the prior exercises, the NASA personnel were assembled and asked to make judgments about the following.

1. Determine the main IFs affecting reliability
2. Rank the importance of the influence factors
3. Define the quality descriptors for each IF
4. Define with the HRA expert the representative scenario categories and develop exemplar accident scenarios for each category

Some variation was seen in their responses.

10.4.4 SELECTION OF INFLUENCE FACTORS AFFECTING CREW PERFORMANCE

Visits were made to the site and involved witnessing scenarios, carrying out post-scenario interviews, and eliciting experts' opinions, and later carried detailed reviews with the persons involved in prior incidents. This later topic was added to gain insight into how effective personnel were responding to real incidents.

Simulator exercises were observed, the elucidation sessions were carried out on subsequent days, and the incident reviews were given on the last day. In the case of NASA study, the total number of persons who took part in the elucidation periods was 33, a group including astronauts, controllers, and instructors.

As an example of the process, during the interviews, the interviewees brainstormed to develop the influence factors. Some 43 IFs were identified, and these were ultimately reduced to six by a process of interaction with the interviewees. A detailed review of the 43 led to the following: some IFs were found to be influences that affected individuals only rather than being systematic influences appropriate to an HRA; some IFs could be interpreted as adding a dimension to one of the key IFs and as such were subsumed in with the key IF; and others were categories of factors that could be broken down into the six key influence factors.

Within the HDT process, IFs have two attributes. For each scenario type, the first attribute is the relative importance or ratings against each other, and the second attribute are the QFs of each influence factor.

Subsequently, the definitions of each of the six IFs were reviewed by a group of top-level NASA personnel and modified to make them clearer to NASA personnel. This was to ensure that NASA understood the basis for going forward with a list of IFs and definitions to determine the importance weights of the IFs. A similar process ought to be carried out with operations staff when undertaking PRA/HRA studies for NPPs, chemical and refinery plants, and other high-risk operations.

The six IFs are similar to those that would be present in main control room (MCR) operations of a nuclear power station. The words used to describe the activities of the MCR supervisor and shift technical advisor or engineer might be different than those used for the flight director (i.e., command, control, and decision making [CC&DM]). However, the approach used would be similar.

The six IFs determined and agreed upon are as follows:

1. Quality of Communications
   - Communications are the transmittal of messages and information between flight controllers and astronauts. The quality of communications is defined by the quality of the protocol and its execution between the controllers and astronauts. The quality relates to how information is transmitted and confirmed so that problems can be solved in a timely manner. The quality depends on the roles of the participants and how they fulfill those roles to ensure the minimum of confusion during high-stress conditions. Responses to needs must be given in time and with accuracy.
   - The quality of communications depends on a well-designed protocol, a disciplined and trained use of the protocol, and experienced and operationally tested personnel using it. The quality of the communications is high when few errors are introduced during operations.
   - Communication IFs may range from supportive to adequate to adverse. There are two aspects associated with communications; these are selection of a basic protocol for good communications and adherence to that protocol.
2. Quality of Man–Machine Interface
   - Man–machine interface is the mechanism by which information about processes is accessed by humans and how humans can effect changes in systems by their actions. The information is usually obtained from visual display units (VDUs) and actions taken by the operators are by manual and computer-based commands, which can be hard or soft (computer generated). One thing that should be pointed out is that at the time of study, the computers in the ISS were laptops of an early vintage, whereas the controllers' computers were quite advanced, so there was a difference in the MMI capabilities between the two installations at the time of the study. It is difficult to rapidly replace ISS equipment because of the need to ensure the quality of the equipment for space service.

- The quality of the MMI depends on the organization of information, speed of access of the information, and the ready absorption of the information by the users to enhance their performance in responding to accidents and incidents. As mentioned above, the displays presented to both controllers and crew should meet good human factors principles, the computer architecture should enable an operator quick access to relevant data without large mental effort, displays should be available within seconds from being selected, and the format of the displays should enable the user to quickly identify problem areas.
- The quality of the MMI IF can be divided into three ranges: supportive, adequate, and adverse. A supportive MMI would fulfill all of the above. Adequate MMI might be deficient in one or more areas, but the system can be used to achieve good results. An adverse MMI leads to many unacceptable results, and it is difficult to obtain consistently good results.
- Sometimes the human–system interface (HSI), and human–computer interface, or some other expression may be used instead of MMI.

3. Quality of Procedures
- Procedures are a series of instructions in book form, in flow sheet form, or are computer generated so as to assist operators in responding to both normal and abnormal events. They include normal, abnormal, and emergency procedures as well as other management rules, schematics, and other documents as well as procedures.
- Procedures should be both technically accurate and easy to understand and use. The procedures should be human factor designed. The procedures should be validated to ensure technical accuracy and should be designed for easy use by personnel without undue training. Good procedures should lead the user to the correct response without confusing the user. An important feature is that the procedures have the same "look and feel" in going from one procedure to another.
- The procedures come in a variety of qualities from supportive to adequate to adverse. The supportive procedure meets the entire requirement, adequate means that the procedures meet many but not all of the requirements, and an adverse procedure may have difficulty meeting all of them.

4. Quality of Training
- Training is a process to transfer knowledge to others in an organized manner. There are several ways to do this, ranging from lectures given by knowledgeable persons to "on-the-job training," to the use of simulators to learn by doing.
- The function of training is to prepare various personnel to perform their functions correctly and in a timely manner. Training improves a person's knowledge about systems, how they function, and their relationship to other systems. Training also improves the understanding of the personnel in the overall objectives of the operating the equipment economically and safely.

- The quality of training depends on a well-designed plan for training, the availability of experienced and qualified instructors, exposure to simulation methods to ensure exposure to as near to real-life experiences in both normal and abnormal operation, and a system to check the performance of the personnel and use information derived from the tests to correct and enhance personnel training. There should be a systematic approach to training to ensure that all personnel are well prepared to deal with emergencies using procedures and deep knowledge of systems and operation of the plant.
- The quality of training can be judged to be good, adequate, or inadequate. The differences between these states can relate to how well the astronauts and controllers perform.

5. Quality of CC&DM
   - Command, control, and decision making make up the process of organizing the response of the astronauts and flight control staff to both normal and abnormal events. In this context, it is being considered only for abnormal events. The term *protocol* is shorthand to describe the set of customs, practices, and procedures regarding the hierarchy of responsibilities associated with CC&DM during an abnormal situation.
   - The quality of the CC&DM influence function depends on the organization of the astronauts and flight controllers. The correct functioning of all of the astronauts and flight controllers can lead to an operation that can be classified as excellent. The role of either the flight controllers or the astronauts is central to supplying the flight controller with information and guidance to make the correct decisions to combat accidents and take the steps to recover the systems. The balance between the roles of the personnel depends on the accident. Quality of CC&DM can be judged to be efficient, adequate, and deficient.

6. Degree of Workload
   - Workload is a series of actions to be taken by a person or team in responding to an event in a given time. The perception is that a large workload leads to increased stress, but often stress is induced in the operating persons by their perception that they are not in control of an abnormal event. Workload is usually associated with both the number and simultaneousness of tasks to be performed.
   - Workload stems from the tasks associated with the accident sequence and from organizational requirements. The organizational requirements can impose workloads that are independent of the actual accident sequence and thus can affect overall team performance. From past observations of operators, it was observed that pressure increases upon personnel when they do not think that they are in control of an accident. Under these circumstances, the capability to plan and organize is impacted, leading to less efficient use of time, and this appears as an increase in the workload. This can be seen by others as a perceived lack of time to carry out the work. Workload can be judged to be less than capacity, matching capacity, and more than capacity.

The Project Plan of Figure 10.3 is a diagrammatic representation for the development of an HDT model and its quantification for a PRA. The HRA specialists should develop guidance about the process, and the plant personnel should carry out the process. In the case of the ISS, NASA personnel supplied much of the expertise for the construction of the HDT structure and HEP estimations.

10.4.5 DEVELOPING IF IMPORTANCE RANKINGS

Following the selection of the influence factors, the next process is to rank them. The AHP (Saaty, 1980) is a way to determine the importance ranking of each influence factor. The AHP uses paired comparisons of the relative importance of each factor to construct symmetric matrices. It was proven (ANP) that the relative ranking of the factors that are being compared equals the normalized eigenvector of the largest eigenvalue of the matrix. To carry out the approach in the NASA example, sheets were developed, an example of which is Figure 10.4, which were to be filled in by NASA personnel. In each cell of the sheet, the individual's judgment was entered about how more or less important each influence factor was when compared to each of the others. This is the notion of paired comparisons. This process should be done for each scenario.

However, it is very time consuming to do this for all PRA accident scenarios. To enable the PRA staff to expedite the process, the scenarios in the PRA should be grouped to reflect approximately similar characteristics as far as influences that affect the flight controllers or astronauts. One could, to the first degree, assume that the flight controllers or astronauts are affected similarly for all such scenarios. In the case of the ISS PRA study, it was argued that there were five different scenarios. Depending on the scenario, flight controllers or astronauts may be separately

INFLUENCE FACTOR RATING MATRIX
For Scenario #1

Check One: Crew___ Controller ✗ Instructor___	Scenario Description: Scenario category (1) Crew is dominant in executing procedures. Crew rely on onboard computers and have a great deal of communication with the controllers.					
	Communications	Procedures	Training	Man-Machine Interface	Command Control & Decision-making	Workload
Communications		3 / 6	8 / 6	7 / 6	5 / 6	4 / 6
Procedures			8 / 3	7 / 3	5 / 3	4 / 3
Training				7 / 8	5 / 8	4 / 8
Man-Machine Interface					5 / 7	4 / 7
Command Control & Decision-making						4 / 5
Workload						

FIGURE 10.4 Typical paired comparison sheet from Expert Elicitation of influence factor importance ratings.

		CC&D	Comm	MM inter	Procedure	Training	Work Load
Scenario 1	Total Mean	0.17	0.12	0.22	0.12	0.23	0.14
Docking	Std. Dev.	0.03	0.04	0.04	0.06	0.02	0.04
Scenario 2	Total Mean	0.17	0.17	0.14	0.18	0.21	0.13
Fire	Std. Dev.	0.04	0.04	0.04	0.04	0.02	0.05
Scenario 3	Total Mean	0.14	0.19	0.17	0.21	0.16	0.13
Coolant Leak	Std. Dev.	0.03	0.05	0.04	0.05	0.03	0.05
Scenario 4	Total Mean	0.17	0.24	0.13	0.19	0.15	0.11
Loss of	Std. Dev.	0.07	0.06	0.06	0.06	0.05	0.05
C&C DM							
Scenario 5	Total Mean	0.13	0.2	0.15	0.16	0.19	0.17
EVA	Std. Dev.	0.04	0.03	0.03	0.03	0.03	0.03

FIGURE 10.5 Summary of influence factor importance weights for five scenarios.

affected, and in other cases, both may be affected. The scenario determines which group is affected and how.

The five scenarios selected were docking, fire, coolant leak, loss CC&DM, and extra vehicular activity (astronauts dress in space suits and work outside the ISS). Docking can be both manually and automatically controlled. The selection of these scenarios goes deep into the operation of the ISS and is not very relevant here to the general purpose of the example. However, it should be pointed out that some things do not operate the same in space as on the ground; for example, the gravitational forces are small and therefore fires do not behave the same way.

For the ISS study, there were 15 participants, each of whom submitted five such comparison sheets, one for each representative scenario. Expert Choice software was used to evaluate each sheet (Figure 10.4) from which to derive the importance rating of the influence factors. For each of the five scenarios, the mean and standard deviation of the ratings (of each influence factor) by treating each of the 15 personnel as an equally weighted data point. Figure 10.5 shows the results for the five scenarios.

Before the above process was carried out, a package of information was assembled and handed out to the participants for the ranking and rating task. The package contained the list of IFs and their definitions, details on the paired-comparison method, and a brief definition of the selected scenarios. The participants were then invited to read the materials and ask questions before filling out the forms.

Before the participants were asked to fill out the form for a given scenario, the HRA expert presented details about the expected crew response and included references to procedures that might be used. This process helped each of the participants to have a clear idea of the activities of the flight controllers and astronauts during each accident and augmented the documented materials. The participants were asked to focus on the relative importance of each of the IFs and not how effective they might be in practice. Figure 10.4 shows a typical paired comparison table filled out by one of the participants of the NASA study.

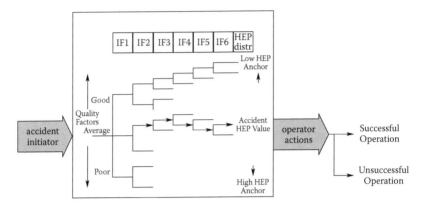

FIGURE 10.6 Pathway through holistic decision tree depicting the effect of variations in influence factor quality.

10.4.6 HOLISTIC DECISION TREE CONSTRUCTION

The elements required for the construction of an HDT model are a list of key IFs, the weights of the IFs relative to each other, a list of quality descriptors, the quality values, and the anchor values. A portion of an HDT developed for this analysis is shown in Figure 10.2. It is assumed that the IFs are independent, but we know in the general case they need not be. This means that the arrangement of IFs on this event tree can be arbitrary. The descriptors or branches cover the range of values from good to poor or equivalent statements. The next section discusses the descriptors in more detail. Once the tree is constructed, the HEPs are calculated from the relationship between the IF importance weights, quality values, and anchor values. Examination of the HDT shown in Figure 10.2 reveals the low HEP at the top end and increasingly higher values (but not monotonically so) toward the other end. Typically, we specify the upper and lower bounds corresponding to the best and worst combinations of IF quality descriptors by a combination of data and judgment. These are considered anchor values. Figure 10.6 shows a pathway through an HDT showing the impact of variations in the quality factors (QFs) associated with each IF. The figure shows only ternary variations in QFs. In practice, the variation of QFs may go from binary to higher breakdowns. The QF groups depend on the confidence in the analysts' representations and depend on the availability of data or if the ranges are supported by expert judgment.

10.4.7 SELECTION OF QUALITY DESCRIPTORS AND RATING THE INFLUENCE FACTORS

As stated above, the pathway through the decision tree depends on the range of the QFs available to the analyst. In the approach to the formulation of the HDT method, the stance taken has been to try to draw upon established technologies to aid in determining the definition of the QF levels. For example, human factors documents used by government and other organizations have given specific guidance to improve MMI (HSI) implementations, and HDT draws upon this work to try to establish the various QF levels. A similar approach has been taken with the QFs for the other IFs. It appears that some of the QFs can be defended by the use of expert judgment. It is not totally clear that HF rules can be supported by experimentation, but they are often supportable by the opinions of experts who have both applied and seen the result. However, humans can adapt to changes with experience and training, so some HF improvements may be mythical. In general, having a suitable framework is a good starting point to provide a breakdown of QFs.

The following is an example of the explicit description of the QFs for the influence factor: CC&DM.

Efficient

Collaboration among team members, roles and responsibilities, and decision making are needed to identify the problems and perform all required tasks encouraging an expeditious return to the desired state of the system.

Example features of efficient CC&DM include the following:

1. Well-defined lines of authority
2. Established protocol and experienced leadership
3. Experienced team used to working together to solve problems quickly

Adequate

Collaboration among team members, roles and responsibilities, and decision making are needed to identify the problems and perform all required tasks neither encourages nor interferes with the expeditious return to the desired state of the system.

Examples of adequate CC&DM include the following:

1. Defined lines of authority, but some uncertainty
2. Established protocol, but not experienced leadership
3. Some experienced personnel used to working together

Deficient

Collaboration between members interferes with the ability to resolve problems and return to a desired state of the system.

Example features of deficient CC&DM include the following:

1. No clear lines of authority
2. Inexperienced team members or ambiguous protocol
3. Inexperienced leadership

The original set of QFs included four different descriptions, such as efficient, adequate, fair, and deficient. However, on review of the QFs, it was seen that it would be difficult to differentiate between the middle two QFs. Therefore, all influence factors are rated using three QFs. Table 10.2 illustrates a set of typical quality factor descriptions for five scenarios (ISS-HRA/PRA Study).

10.4.8 Selection of Quality Descriptor Values

Given the quality descriptors for each influence factor, the HDT method requires an analysis of how much change in reliability is associated with a change in quality

TABLE 10.2

Listing of Estimated Quality Descriptors for the Five Scenarios

Influence Factor	Man–Machine Interface (MMI)	Procedures	CC&DM	Training	Communications	Workload
SC1 (Docking)	Adverse	Adequate	Efficient	Adequate	Supportive	More than capacity
SC2 (Fire)	Adequate	Supportive	Adequate	Good	Adequate	More than capacity
SC3 (Leak)	Adequate	Adequate	Adequate	Adequate	Supportive	Matching capacity
SC4 (CC&DM)	Adverse	Adverse	Efficient	Adequate	Supportive	More than capacity
SC5 (EVA)	Adequate	Adverse	Efficient	Good	Adequate	Matching capacity

(e.g., how much would an HEP change if the quality of procedures changed from supportive to adequate). Because one is interested only in ordinal increases, a QF value of unity was assigned for each influence factor at its best rating (e.g., good, efficient, or supportive).

Lower ratings would increase HEP, and this was reflected by a QD value greater than one. As part of the plan for building the HDT model for an NPP application, one would carry out extensive elucidation similar to that carried out for NASA.

As it turns out, the AHP also uses an ordinal scale to compare levels of preference, importance, or "goodness." AHP uses a 1 to 9 scale. We selected a factor of 3 to represent the transitions from good to fair and fair to poor. Thus, the HEP values were generated using a quality descriptor value of 1 for Good, 3 for Fair, and 9 for Poor rating of the influence factor.

10.4.9 SELECTION OF ANCHOR VALUES

The selection of the upper anchor value is based on the certainty that if all aspects that affect the performance of humans are bad, then it is very likely that they will fail. If the effect of the context is such that humans are not given any help, then it is certain that they will fail, and the probability of failure will be one (1.0).

The lower anchor value can vary, and uncertainty in the lower value can be included in this analysis. The lower anchor point corresponds to the context being the most favorable in that all of the IFs are at their best values. In the ISS case, one can expect that the crew or Mission Control Center (MCC) will have a high probability of success. Each scenario has been assigned a different distribution for the lower anchor value. For most scenarios, the 5 percentile of the low anchor value is set at 1.0E-04 and the 95 percentile at 1.0E-3. For some mild cases, the lower bound is set as 5 percentile at 1.0E-3 to 95 percentile at 1.0E-02; and for severe cases, the lower bound is set at 1.0E-02 for the 5 percentile value and 0.1 for the 95 percentile value. These judgments reflect the fact that in most cases, there is time available for the personnel to reach a resolution and correct the situation.

10.4.10 CALCULATION OF HEP VALUES

An Excel worksheet was developed to calculate the distribution of HEPs given the selection of IFs, weighting values associated with the IFs, QDs and their values, and the selection of anchor values. For this worksheet, the number of IFs is six, and the number of QDs is three. The program currently includes the ability to quantify uncertainty in each IF rating (weighting values) and the uncertainty in the lower- and upper-bound anchor values. Between the upper and lower anchor values (i.e., the first and last pathway), a logarithmic interpolation is used as follows:

$$\ln(HEP_i) = \ln(HEP_l) + \ln(HEP_h / HEP_l)\left[\frac{S_i - S_l}{S_h - S_l}\right] \tag{10.1}$$

$$S_i = \sum_{j=1}^{n}(QV_j)I_j \text{ with } \sum_{j=1}^{n}I_j = 1 \tag{10.2}$$

where HEP_i is the human error probability of the ith pathway through the HDT; HEP_l is the low HEP anchor value; HEP_h is the high HEP anchor value; S_l is the lowest possible value of S_i. This equals 1.0 in the current formulation. S_h is the highest possible value of S_i. This equals 9.0 in the current formulation. QV_j is the quality descriptor value (i.e., 1, 3, or 9) corresponding to the jth influence factor. I_j is the importance weight of the jth influence factor as determined from the NASA expert elicitation process and evaluated using the AHP. n is the number of influence factors in the HDT. The numbers of influence factors are six in this case.

The notion that the logarithm of an HEP is proportional to the weighted sum of influence factors anchored by upper- and lower-bound values was first introduced by Embrey and colleagues (1984).

In Figure 10.2, the numbers that can be altered are related to Quality Factors (QF). Quality Values (QV) are associated with each identified QF and are assigned by the HRA analyst. There are 729 branches altogether corresponding to six influence factors and three values per influence factor. Each pathway through the HDT is for a different set of six quality descriptors. Table 10.3 lists descriptions for a given pathway associated with a given accident. Because each pathway has a different set of QVs and each influence factor has a different weighting factor, each path has a unique HEP. For example, the last (bottom) pathway in Figure 10.2 corresponds to the following: quality descriptors of influence factors.

This corresponds to $S_i = 2.56$ and $HEP_i = 1.18E-03$. In order to determine the HEP for a scenario in the ISS PRA, one finds the pathway that corresponds to the set of quality descriptors that were found for the overall scenario.

TABLE 10.3

Listing of Influence Factors for a Given Holistic Decision Tree Pathway

Influence Factor	Quality Descriptor Rating
Communications	Good
Procedures	Good
Training	Poor
Man–machine interface	Good
Workload	Good
Command, control, and decision making	Fair

10.4.11 HEP RESULTS FOR A REPRESENTATIVE ACCIDENT SCENARIO

Evaluation of the HEP using the HDT method described herein produced the following results for the representative accident scenario. Table 10.4 provides summary statistics for one scenario, and Figure 10.7 is the HEP probability distribution.

The results are sensitive to the anchor values. The results are less sensitive to the individual influence factor importance weights, because no single influence factor is rated as overwhelmingly important relative to the others.

Uncertainties not quantitatively included are as follows:

1. Influence of a future data mining and analysis effort that would provide data-driven anchor values
2. Selection of alternative definitions of the influence factors, quality descriptors, and quality values
3. Selection of different quality value increments for the quality descriptors
4. Alternative influence factor ratings

TABLE 10.4

Statistics for Human Error Probabilities for the Scenario

Scenario	SC 1: Docking
Mean	4.6E-03
Standard deviation	2.1E-03
Variance	4.4E-06
Mode	4.3-03
5th percentile	1.9E-03
95th percentile	8.4E-03

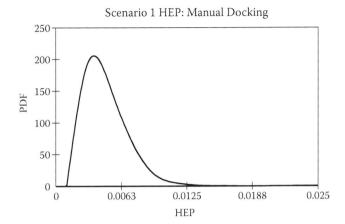

FIGURE 10.7   Plot of human error probability variation for SC 1, manual docking.

10.5   COMMENTS

The HDT method was selected to meet the requirements of the independent ISS PRA review team. The project was a start in the process of the incorporation of NASA experience and knowledge about space operations into the risk assessment environment of the ISS. Unfortunately, two events conspired to ensure that the HDT method is not currently being employed by NASA.

The final stages of the project were held up by the effects of the Columbia accident on the NASA organization. Work on the update of the HRA technology was in progress from November 2003 to July 2004. The Columbia accident took place February 14, 2004. For some time, the HRA project was held up, so for the critical last phases of the project, access to NASA personnel was restricted. The use of NASA personnel to cover the last phases of the project was restricted, and therefore, the HRA analyst had to complete the QF estimation and other aspects without the use of NASA personnel.

It was hoped that NASA would fund the completion of the study and the application of the HRA method to more scenarios, but this was not to be. A change in the support contractors to NASA came about, and this meant that the original PRA/HRA personnel were replaced by other contract personnel.

Nonetheless, it is believed that the advances in HRA technology could still be of use to NASA. The experiences undertaken to inform and train NASA personnel in HRA methods and procedures should be of help to NASA personnel to appreciate risk assessment and the importance of HRA as a risk contributor and something of the processes used in HRA field, especially as far as expert judgment and elucidation processes are concerned.

The steps in the HDT process outlined here can be used in other PRA/PSA applications and are pretty standard. The split between astronauts and controllers, as far as accident mitigation is concerned, is unlikely to occur in other applications, but the process steps to determine the IFs, QFs, and anchor values could be used. Expert elucidation is fundamental to HRA, because one wants to capture the specifics of the actual situation rather than just a derivative of a generic case.

The application of the HDT method shows that it is a very useful concept that ties together all of the context elements to produce a result that reflects the response of the operators, whether they are astronauts, controllers, or nuclear control room operators. The impact of changes in context is important for managers to be aware of what can lead to reductions in risk. It then can put them in a position to estimate the risk–benefit of any proposed modification and pick the best option, in terms of cost and schedule.

11 Data Sources, Data Banks, and Expert Judgment

INTRODUCTION

Although many people have pointed out how valuable it is to draw conclusions from qualitative insights about human activities, it is still necessary to establish values about the relative risks related to given situations. This is what one can achieve by the use of numerical probabilities for human actions within the logical framework of the PRA/PSA.

To establish numerical values, one needs to have access to data relative to the actions. This chapter considers the availability and use of a variety of data sources. Some of the data are given in databases associated with particular HRA methods, such as THERP (Swain and Guttman, 1984). Other data exist in the form of data banks, such as CORE-DATA (Gibson et al., 1999). Another source of data is found by the use of expert judgment from known domain experts. Another interesting source of data is from simulator records. Chapter 12 talks about the use of simulators for HRA purposes. Data can be collected by the use of simulators, but there are some limits to its usage, and these will be discussed in this chapter.

Most plants and regulatory authorities collect data on safety-related plant accidents and incidents. The U.S. Nuclear Regulatory Commission (USNRC) collects what are called licensee events reports (LERs). The Institute of Nuclear Power Operations (INPO), and its associated world organization, World Organization of Nuclear Operators (WANO), also collects information; unfortunately access to these databases is restricted. One interesting activity is that of the French organization Institute for Nuclear Reactor Safety (IRSN). Apart from their other work, they have being collecting data on plant operations that lead to operational problems other than safety-related incidents. The program that embodies these records is called Recuperare (Matahri and Baumont, 2008), and this should yield some interesting insights into incidents that could lead to accidents and the interactions between management and operations. The French along with others also collect accident-related data.

There is a history of collecting human error–related information in the form of a database; there were a number of early databases and some early work on human reliability, see the references at the end of this chapter. The USNRC awarded a contract to General Physics to develop a data methodology in the 1980s for a human database (Comer et al., 1983). The ownership of the database passed to Idaho National Laboratory and became NUCLARR (Gertman et al., 1988). The premise for the early version of the database seemed to be more associated with different plant types and components, rather than focusing on the tasks and the conditions under which the errors occurred. Individuals were asked to contribute to the database, but

NUCLARR seems to have disappeared from the HRA scene. What could be thought of as a replacement is the human event repository and analysis (HERA) NUREG/CR-6903, Volumes 1 and 2 (Hallbert et al., 2007). It was seen by a number of persons that there was a need for a useful database covering human reliability aspects. HERA has evolved over many years from its first concepts. Its origin can be traced to a need identified in the mid-1990s. The specific purpose of HERA is to analyze and store human performance information from NPPs operating experience that is of regulatory interest. HERA should supply data to lead to a better understanding of human reliability in an operational context. This knowledge should lead to improvements in both HRA methods and from the point of view of the USNRC support for regulatory applications.

There are issues associated with every data source, and these will be discussed here. Questions are raised with respect to the data, its accuracy, and its relationship to the current evaluated situation. The questions are as follows:

1. How are the data obtained?
2. How are the data from different data sets treated?
3. How relevant are data to the current application situation being evaluated?
4. Are the results accurate, and can you trace the data back to its sources?
5. Has it been peer reviewed?
6. Is the database flexible enough to be used with different HRA models?

There are a number of different data sources, some associated with a given HRA model, and some of the databases are free standing and have been used by some investigators and others who draw upon the use of expert judgment or domain expertise to provide HEP estimates. Other methods are based upon knowledge of how the context of an accident sets up the probability of human error. It seems that there is a compulsion for some HRA model constructors to develop not only a method, but also include an associated database as part of their work. Sometimes this works out quite well in that it fulfills a need for persons tasked to carry out a study, but who do not have access to plant-related data, for example, if one was to estimate the operational risk of a yet-to-be-constructed NPP.

This chapter discusses the various available data:

1. Associated databases
2. Stand-alone databases
3. Simulator databases
4. Expert or domain-generated databases

Associated databases are those that are part of the documentation and approach associated with a given HRA model. A stand-alone database can be used by any investigator and does not need to be tied directly to a given HRA model. Simulator databases reflect data and information that can be collected. These databases can be interrogated for information on how individuals and crews respond to simulated accidents and record the corresponding time data related to the response of the NPP. The information can also be used to compare the responses of the crews.

The last category is the expert judgment database. Expert judgment can be used during the application of HRA studies in a piecemeal manner to satisfy an immediate need to estimate the impact of a PSF on an HEP, for example, in the application of NARA. But expert judgment could be used to estimate HEPs for given human situations in accident scenarios, and this data could be collected and stored for later use. Mostly simulators do not yield statistically significant error rates; therefore, there is a need to use expert judgment to estimate HEPs. The opinions of the experts can be augmented by reviewing crew performance based on the simulator-related information.

11.1 ASSOCIATED DATABASES

Associated databases are those that form part of a developed HRA method. In part, they consist of human error probability (HEP) distributions related to a given task description. In addition, the database relates to various PSFs, EPCs, and CPCs. Also, the method contains guidance on the balance between these performance factors. These modifiers are called, for example, assessed proportions of affect (APOAs). This type of database is seen in HEART, NARA, and CREAM; see Chapter 5.

Technique for Human Error Rate Prediction (THERP) focuses on the immediate data contained in the various parts of Table 20 within the handbook (Swain and Guttman, 1983). The PSFs in THERP are a little vague in how to treat them. Mostly it deals with stress, and one must disagree with the use of high stress associated with the large break loss of coolant accident (LBLOCA), according to some operators, because there is very little that they can do, they say that there the stress is low. From observations of crews tackling various accidents at a number of power plant simulators, it is the opinion of the author that crews feel a similar degree of stress for all accidents apart from the trivial. Once the accident is understood and under control, the stress level felt by the crew drops.

The questions raised about databases are as stated in the introduction to this chapter. It appears that British Energy is not too sure about the HEART database, so one would defer to them here. The origin and support for THERP are also tenuous. Part of the problem might be a lack of either transparency or the availability of appropriate documentation. One should be aware that the NPP industry has changed remarkably since the early 1980s; so even if the data in the handbook were defendable at the time of being published, the industry moved so far as to make that data somewhat incredible if not irrelevant in today's world.

The steps taken by the NARA developers appear to be in the right direction in using an outside database. One still has concerns about the database (CORE-DATA; Gibson et al., 1999). A very limited review of the data indicates a lack of enough data to enable a statistical support for some of the HEPs and their distribution. One should be very careful how one uses HEPs and their distributions, considering limitations within the data set. The other issue is the same as that raised for THERP—how relevant is the CORE-DATA to the current situation? If the database is not fully transparent for the knowledge expert to evaluate, should he recommend its use? Another issue arises in that all three methods, HEART, NARA, and CREAM, use field data to some degree. The model method then corrects for the HEP distribution by loading it with correction factors of one kind or another, so should one apply a correction factor to the input data to ensure that it is "clean" data representing a sort of ideal or "platonic" set of data?

So one can see a number of concerns about data, in terms of where does it come from, do we know the conditions surrounding its collection and interpretation, and are the PSFs associated with the data known and compensated for? The pedigree of data is important. Is the data a mongrel or a pedigree with a certificate? One places a lot of faith in the results of the PRA. Some uncertainty in the data is expected, but if the uncertainty is too high, can we draw the right conclusions from the results or are they meaningful?

11.2   STANDALONE DATABASES

There are pros and cons to using all forms of databases. For the HRA user, it is very comfortable to pick up an HRA model or method that has all of the available parts to be able to generate an HEP value to enter into the PRA. For all of the alternative databases, one has to go outside of one's comfort zone. The developer of an associated database HRA method has supplied all of the elements for an HRA analyst to be able to fulfill his job. Also, if the method has the approval of the management and the regulatory authority, one does not have to prove that it is the best. In fact, the concept of consistency seems to be more important to some authorities than accuracy. This choice should be questioned, because it could lead to a situation in which all are unsafe NPPs to some degree or in some situations.

However, if there are questions raised about the appropriateness of the data, what are you to do? One thing to do is to accept the model or method but not the data, because it is clear that if the original database was constructed from data from another country or place or time it cannot be fully defended. One may also not have sufficient information about the data sources for the model's associated database to be able to defend the results. Under these circumstances, one looks for a database that is current, available, well documented, and peer reviewed. The NRC is funding a project called the human event repository and analysis (HERA) that appears to meet these requirements. Another independent database is CORE-DATA. Both of these databases seem to fit the bill for an independent database.

HERA is a framework for describing operational events involving human performance. It includes a method for analyzing the events and extracting relevant data about an event. The software enables one to view the contents and store the data.

HERA has one big advantage in that the USNRC has an open policy as far as reports are concerned. However, there may be some limits in that some organizations that enter data into the database may also want to apply access restrictions to their data. This would reduce the value of such a database. In the past, some utilities allowed simulator data to be used provided that the data were not directly linked to the NPP control room crew members. In dealing with a database, one is interested in what situations lead to human errors, not in assigning blame to crew members.

The objective of HERA is to make available empirical and experimental human performance data, from commercial NPPs and other related technologies, in a content and format suitable to HRA and human factors practitioners. Provided that the data entered into the database have a pedigree, all should be well. The data should be traceable to a real situation and type of plant, and simulator data should be included. It may not be necessary to identify either actual plant for actual plant data. The same is true about the names of the MCR crew, their names and positions, and any other operation personnel.

The current status of HERA (Richards, 2009) is that the database contains some events that have occurred at U.S. NPPs. The database at this moment has entered a sufficient number of events to ensure that the software package works well. It has been pointed out that the records from events can contain anywhere from 50 to 150 subevents. Clearly, some events involve more human actions than others, and this is reflected in defining the number of subevents. Reviewing accidents and trying to capture relevant information and entering the data into the database can be labor intensive depending on source materials. Data to populate HERA have at this time come from publicly available reports, like Licensee Event Reports (LERs). Trying to capture useful data from events is difficult, as can be seen by the difficulties encountered in Chapter 9, while trying to analyze human reliability events from accident reports. HERA has been produced and reviewed under the typical way that the USNRC considers projects and documents. This is not quite the same as an independent review, because the reviewers are associated with the USNRC, but by any normal measures this is a good process. Documents are also issued for public comment, before being released as official documents.

CORE-DATA seems to be a useful database, but data contained within the database should have a credible pedigree, and the data process should be transparent. Here the report should cover details of the data, such as where did it come from, which industry, and how was it used to calculate HEPs, if that was the objective of the database. One should also be able to enter the database and construct other output forms to use with other HRA models.

It is believed that when NARA was peer reviewed by a group of international experts, CORE-DATA was also part of that review. The author has no personal knowledge of the internal data within CORE-DATA, so he cannot attest one way or another that all is acceptable. Since the program was started at the University of Birmingham, one would have thought that it would have been freely published.

11.3 SIMULATOR DATABASES

As mentioned in Chapter 12, there are data collection systems available to collect valuable simulator data for HRA, human factors, and training purposes. Unfortunately, the industry is losing by not collecting and analyzing this valuable data. Some individual utilities have collected simulator data and subsequently used it for purposes of defending utility decisions relative to concerns that the USNRC had with respect to the reliability of certain operator actions. An advantage in simulator data, especially for the plant involved, is all or most of the PSFs or compensating influences are already covered. To make simulator data available for other plants, one needs to be certain of what influences are controlling and try compensating for them.

In comparing U.S. operator performance with Taiwan operator performance, one needs to know the similarities and differences in both the plant design and those issues that affect the operators, such as training and procedures. In one case examined, the dominating influence for the ATWS accident was the design of how to achieve the requirement of dropping the control rods into the reactor core given the decision that they should be. The results for the two PWRs were similar, but the time scale was different. Cultural differences between the U.S. and Taiwan crews were difficult to see. The language was different, but the trained responses of the crews were very close. Both sets of crews used symptom-based procedures, the HSI design was similar, and the training with respect to the importance of the ATWS accident was the same. In countries where the underlying philosophy as far as procedures are concerned, the results are different and one should not use HRA data without understanding the operating philosophy. Hence, it is hard to justify using a specific method that is acceptable for one operational context but is unacceptable in another context.

There were a number of simulator data collection projects in the United States, France, Hungary, Czech Republic, Japan, China, and Korea that have been based on running accident scenarios in order to use the data for a number of different purposes, from formulating rules for governing, whether an action could be safely performed by humans, to investigations related to HRA applications, to the validation of EOPs, and to the evaluation of operator aids of one kind or another. Data exist in various reports, some by EPRI, NRC, and VEIKI/Paks. Currently, the USNRC is engaged with Halden Laboratory in exercises to collect and analyze data for various accident scenarios to help in the evaluation of various HRA models. Here the object is to help the USNRC to decide which HRA model or method best fits USNRC requirements. The USNRC sponsored a report, NUREG-1842 (Forster et al., 2006), "Evaluation of Human Reliability Analysis Methods against Good Practice." However, beyond the Electricité de France simulator-related database, there is no established database for a number of plants. It could be that data from these various studies are held in some data bank. EPRI and the U.S. DOE did at one time sponsor programs to develop methods for the collection of comprehensive data from simulators, as mentioned above, but no steps were taken to establish a comprehensive database for that purpose.

The USNRC sponsored work undertaken by Pacific Northwest National Laboraories (PNNL) and General Physics to investigate NPP team skills based on observations of operators taking actions in response to accident scenarios. The data taken during these experiments could have formed the foundation for a database of simulator session results. An official report on the results and findings has never been published. It is believed that the draft report (Montgomery et al., 1992) is available in the USNRC document room. The best hope for the future for the collection of this type of data is the HERA database being developed; see Section 11.5. The problem in going back to insert past simulator data is that some of the details that were available at the time the data were collected have not been retained, and this makes the data much less useful than it could be.

As mentioned in Chapter 12, data collection methods have changed from just recording the times when operators took certain actions to more sophisticated data collection systems in which operators' actions, observations of their interactions with other crew members, procedures, and the MMI were recorded along with video records of the operations as backups. This was accomplished in various stages under a series of projects sponsored by EPRI in combination with various utilities and DOE/PNNL. The phases of the project had initially started with the idea of automating the methods used during the ORE project, but soon developed into a tool to support instructors evaluating operator performance. The final version in the development process was a program called CREDIT Vr 3.2 (Spurgin et al., 2000). This program consisted of a combination of a way to collect simulator data via a connection and to input observations of crew activities based upon using bar code readers. The observations were made by instructors (or others) using bar coded sheets. There was taxonomy to help ensure that the observational information collected, fit into a form that helped the instructors and provided the ability to analyze the data later. Support programs would construct the observational data in a form to aid the observers, while enabling the analysts to sort the data into useful forms for HRA and other purposes. One problem with instructors recording their observations was the ability to extract useful information over a group of such observations. Their observational comments did not normally fit into a useful protocol from which it was possible to draw long-term conclusions as to the quality of the training program judged by trends in the performance data. Another issue was associated with the idea of exposing the crews, as a group, to a set of random accident scenarios. This selection process makes it extremely difficult to monitor the crew-to-crew variation in accident response.

One should mention the work of EDF and its research personnel. EDF has been collecting simulator data in one form or another since the early 1980s. The personnel involved have changed over the years, and some of the analysis processes have changed. EDF has committed resources to collect data at simulators. The current HRA method is MERMOS (LeBot et al., 2008), and it is based upon collected information and placing the data in a database. In this case, the data or information is collected in a Lotus Notes database. It is the intent to place the data into new dedicated software that is under construction, called IDAFH. This will structure the MERMOS analyses with a Bayesian network and will record simulator data and events (LeBot, 2008b). The MERMOS data/information is recorded mostly as text and sometimes figures.

In terms of observational data, they tend to use three persons with a background in human factors, so their observational skills should be better than using instructors for this purpose. Analysis of the data for various accident sequences at simulators is used within the MERMOS process.

In some ways, EDF is in a good position because they have essentially three NPP designs, so some of the variability that exists elsewhere is not present. It would be interesting to examine results for crews from different stations to see if there are different error rates from plants at different locations. Although the dynamic characteristics of PWRs are similar (all EDF NPPs are PWRs), there are differences between the early three-loop PWRs (900 MWe), the four-loop PWRs (1300 MWe), and the later N4 NPP (1400). The early plants use manually read symptom-based procedures, whereas the N4 plants had a computer-directed method for procedure help for the operators. This later system was one reason for the development of MERMOS.

11.4   EXPERT JUDGMENT METHODS

Expert judgment methods are very much in the center of HRA methods and appear in different guises. Sometimes HEPs are estimated directly, and other times expert judgment is used to select the nearest task to one given a list of tasks that have been identified as canonical tasks. This choice makes available the distribution of the basic data for the situation. However, the HRA model user has the ability to select from a set of PSFs or equivalent, a number that the user feels is significant in shaping the actual probability distribution. Furthermore, the user can, if he or she so wishes, adjust the distribution of PSF contributions, but the selection weighting factors. The relative importance of the PSFs is selected by the designer of the HRA method. In the mind of the HRA designer, there is a distribution of PSF contributions.

The point here is that much of the basis of HRA evaluation is based upon the judgments of experts—the level of expertise is split into two parts. The first type of expert is the technology or knowledge expert, and the second expert is the domain expert. The first expert is expected to understand the whole field of HRA in terms of methods, models, and basis for the databases to be accessed during a study. This expert is expected to understand which methods and models to use under the study circumstances and the justification for the above.

The second expert is expected to understand all aspects of the operation that is the subject of the study. This covers operator training, plant operation, emergency, and abnormal and normal operating procedures. He should also be knowledgeable about plant operating history and how the plant management operates.

As you can see, the requirements of both experts are extensive, and it is unlikely that any two persons could cover all of the above. However, this does not mean that one can do away with not having this knowledge available to the PRA team. It is expected that some studies have been carried out short of satisfying these requirements, but the quality of the study will be impacted to its detriment.

It is to be pointed out that the selection of the HRA model or method is dependent on the availability of data that can be used to fill in the blanks (i.e., selecting the appropriate HEP values). Even if one is applying a comprehensive method, say THERP, one finds that the range of operations identified in the text (tables) do not match the needs of a particular situation. One is then driven to define a new task and an HEP distribution. The domain expert can see the need for a new task as determined in discussions with the domain expert.

Given the domain expert knowledge, the next thing is to get him to estimate the corresponding HEP distribution. However, this presents a problem for the knowledge expert, so it is up to him or her to establish a method for extracting this information from the domain expert. There are a number of ways to achieve this, and there are a number of papers on the topic. For example, Comer et al. (1984) and others funded by the USNRC produced a good introduction to the topic. Hunns (1982) discussed one method called paired comparisons. One can see an application of this approach in Chapter 10, where NASA domain experts were used to classify aspects of the holistic decision tree method for application to the ISS HRA study.

The USNRC has funded work on an advanced HRA method called ATHEANA (Forster et al., 2000) for a number of years. The early version of ATHEANA used the HEART database for its HEP-associated data. Recently, the developers moved away from the HEART database to the use of expert judgment. They identified an elicitation method (Forster et al., 2004) to involve domain experts in the process of estimating HEPs for use in ATHEANA.

An early HRA method called the success likelihood index method (SLIM), developed by Embrey and others (Embrey et al., 1984), was based on the use of domain experts to estimate the relative influence of a set of PSFs. This method was applied both in the United States (as FLIM) and in other countries. A review of how the method was applied showed some significant issues with using expert judgment to estimate certain things that domain experts are not aware of. For example, does a domain person know how a PSF affects human errors? In other words, how does one use domain experts and gain access to knowledge and understanding?

Clearly one has to work in a way to elucidate the limits of the knowledge and not subsume that they fully understand the underlying HRA concepts. This can be one of the advantages of comparing situation A with situation B, like the paired comparison approach. Data collected from the elicitation processes needs to be examined in detail, for differences in the results between different crew/persons can reveal the fact that results are not consistent (i.e., the distribution reflecting the operators' views may be significantly different). One set of results indicates a high likelihood of failure, and the other a high degree of success. Another case may be that one factor is seen to be significant according to one group and insignificant to another group. One must find out why there is a difference. It could be an issue with the description given to groups tht leads to a different interpretation by the operators. Sometimes one needs to discuss the situation with all the groups to ensure that there are no problems of this kind.

Sometimes in getting the operators to carry out a direct estimation of the probability of failure, they may need a tool to relate probabilities to their experiences. Instead of dealing with probabilities, one could deal with failure events within a set of actions, say 1 in 10 rather than a probability of 0.01. It is a small deal, but sometimes one forgets that not everyone really has a feel for a probability number, especially when the numbers are denoted as 1E-5. A useful tool might be a logarithm scale with both types of numbers given; see Figure 11.1

The expert elicitation methods are quite well known, and the objective is to supply enough information about the situation without leading the domain experts to a conclusion. An outline of the principles is given here, but one should review documents that go into more detail, such as the following: Comer et al. (1984), Forster et al. (2004), and others.

Often one has to educate the experts in the process without setting them up to replay your numbers or attitude back to you. The other objective to gaining good results from the experts is to let each expert make his estimate without interference from the others. Once the estimates have been made, they should be examined for any form of inconsistencies. Afterward, one can address the differences with the group as a whole. During the discussion, sometimes the group identifies why certain domain experts have produced outliers. Typically one would like to get an estimate

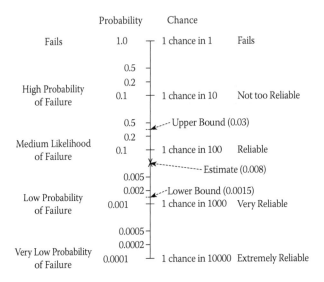

FIGURE 11.1 Expert estimation assistance tool.

of the mean and upper and lower bounds, for example, Figure 11.1 shows a mean of 0.01 and the upper and lower bounds; of 0.03 and 0.0015.

Comer suggested in their document that six or more experts should be used. This is ideal if it can be arranged. In the ISS HRA project, the numbers of experts as a whole were greater than six; however, the group was made up of flight controllers, instructors, and astronauts. However, if one had tried to use just astronauts, the numbers would have been heavily restricted to about two or three, and they would not have necessarily been available for the whole session. Astronauts are always very busy and mostly not very easy to schedule. Other industries may also have a similar kind of problem, in that some of their key experts are not readily available.

In practice, it has been found that a variation in the estimates is to be expected. Wide variations should be investigated, in case they are due to a misunderstanding of the description of an event as presented by the knowledge expert. For example, if the description of an accident is vague, the domain experts mentally build their own versions of the situation, which can give rise to interpretations among the experts. Individually, the accident sequence imagined may be quite different, and hence, the resulting numbers (HEPs) can equally be different. The presence of multiple well-separated peaks in the HEP or PSF distribution represents differences in interpretation of information presented to the domain experts. A balance has to be made between supplying too much information and too little. In the first case, the experts just reflect your opinion; in the second case, the variations in the HEPs may be unacceptable.

Figure 11.1 is useful to assist the domain experts in trying to relate the experiences to the kind of estimate that is needed. Depending on their background, they may feel more comfortable making estimates based on words, ratios, odds, or even probabilities. It is a good idea to start with something that can act as a control, so you can be sure the experts are at ease with the process you are using.

The process can be used for all aspects of the HRA. It can be used to select the nearest task to a description in a table. It can be used to check the base probability distribution against their expectations. In NARA, they could be used to check the description of the task, the base HEP distribution, the relative contribution of the EPCs, and the weighting factors APOAs. Of course, one might not want to do this, because of time and the availability of experts who are really fully involved in operating the plant.

11.5   CONCLUSIONS

The idea of having an HRA database is valuable. The value of a database increases exponentially when its pedigree is known and demonstrated. Its value also increases depending on its ability to absorb various data sources like operational plant, near accident data, and simulator data.

A requirement should be that users have access to either the raw data or processed data. The database ought to be able to provide data for different models. As said above, experts can provide useful HEP estimates. This source of data should also be entered, but lines should be drawn between actual plant data, simulator, and expert judgment estimates. If the user wants to lump the various different data together, the responsibility for that lies with the user. He should provide documentation to show what was done and its justification.

There are a number of different databases, and all offer the user some advantages and some disadvantages. The associated databases are there and can lead to early estimation of HEPs. But these models rely on questionable and possibly outdated data that are not so closely associated with an actual plant.

There is no really usable simulator base at present, although the means to get one was developed some time ago, but the industry moved away from that pathway with the single exception of EDF. EDF pursued the utility of the simulator for HRA and human factors purposes from 1980 to the present. Data from simulators are embodied into the MERMOS HRA method, which is the PSA vehicle for EDF. In this, the EDF has an advantage in that is has an NPP industry that has focused on the PWR and with a consistent policy with regard to the balance of plant (BOP).

The alternative database has a number of advantages in that it should be capable of using plant data, simulator data, and expert judgment data. The HERA database has a number of advantages in that it can contain useful collected data. This data should be capable of being used by HRA developers and also HRA users to fill gaps in their plant data and to give insights into the HRA issues. The HRA user can decide to use HERA to give data that are not available from the built-in associated databases.

Expert judgment is the great supplier of HRA data in one form or another. All HRA methods and models call upon expert judgment even if it is their developer who is doing the calling. The use of expert judgment techniques are referred to and some guidance so as to get the best out of expert judgment is given. The elicitation process requires some skill and practice to be capable of getting useful data. Expert judgment is not like filling the numbers into an equation. The results can vary, and the results cannot be validated. But sometimes it is the best we have. Some PRA scenarios are difficult to test on a simulator, and the opinions of experts may not be that reliable for some conditions. Sometimes the estimations of experts are all that is available.

12 Use of Simulator Data for HRA Purposes

INTRODUCTION

This chapter addresses the use of nuclear power plant (NPP) simulators for human reliability assessment (HRA) purposes as an extension of the normal training of control room personnel. The chapter opens with a brief description about simulator design, followed by a brief history of the simulators and later a discussion of data collection methods for HRA purposes. Their use for training is discussed in Chapter 13.

One of the early uses of simulators for purposes other than training was by the U.S. Nuclear Regulatory Commission (USNRC) in a program aimed at trying to determine where operators could be used successfully for safety-related actions in the early 1980s. A little later a series of data collection projects was undertaken by the Electric Power Research Institute (EPRI), Electricité de France (EDF), and by other countries. The USNRC program was originally designed to support ideas associated with the role of operators performing safety functions.

The results from the USNRC program eventually led to the use of time reliability curves (TRCs) in the HRA field, to the development of the human cognitive reliability (HCR) method, and hence gave rise to various programs to collect and analyze simulator data, including the operator reliability experiments (OREs) funded by the EPRI and others.

Data collection just for HRA purposes is hard to justify, as it can be quite expensive to carry out, when one figures the cost of running the simulator facility and factoring in the cost for instructors and support personnel. HRA data collection can place an additional burden on the instructors collecting data and noting details of operator performance. However, if the needs of HRA are combined with those of training, then the costs are shared; therefore, it is more feasible to collect HRA-related information. It has been found that operator performance data can be a long-term benefit to the operation of the plant and can provide information that can be used successfully to defend plant operations versus regulatory perceptions. This can result in real savings for the power plant.

Many utilities in the United States and other countries use the systematic approach to training (SAT). The real key to the value of the SAT approach is having a good feedback system based on deep insight into human performance. This insight can be obtained by having a well-designed data collection and analysis system, but unfortunately this is not the case for many utilities. It is my belief that one could use a data collection system to provide better feedback information to improve training of the crews, and also provide information to plant management on the quality of operator performance. Often in the current approach to observing operator performance, the

chance for biased observations opens up. To combine both observation and evaluation of operator performance in one act is to ask for trouble. The best is to separate
observation and data collection and then later examine the collected data for the
purpose of performance evaluation. It is the author's opinion that training should be
even more focused on the use of simulators rather than on the use of conventional
classroom activities. This topic is discussed in more depth later in the chapter.

The data collection processes have improved over the years from the early USNRC
approach (Kozinsky et al., 1983), which collected only time data and later spread to
include observer data, time data, and video recordings, such as in the OREs (Spurgin
et al., 1990). The increased emphasis on observer data mirrored to some extent developments in the HRA field. Changes came from a number of researchers, but HRA
researchers were challenged by Dougherty (1990) to produce a second generation
of HRA methods. Erik Hollnagel and P. Marsden (1996) responded by identifying a relationship of context and human error probability (HEP). Others were seeing, through their experiences derived from observations of operator performances
responding to simulated accidents, some of the limitations of the then-current HRA
methods, including THERP (Swain and Guttman, 1983) and the various TRCs.

Simulator studies showed that control room crews sometimes make observable
errors, some of which were recovered. The awareness of the errors became more
apparent once careful observations were made of the operators' performance. Videos
taken during the simulator exercises were very useful in confirming the observations
of researchers. The earlier Oak Ridge and General Physics simulator studies used
time responses of the crews as the only recorded measure of performance and then
focused on successful time responses for the purpose of the study objective. The
errors made by crews did not receive much of an investigation beyond saying that
there were some failures. The basic requirement for these seta of simulator studies was focused on trying to give guidance to the industry on what was the proper
division between man and machine (automated protection systems) in the case of
responding to accidents.

A typical nonsuccess TRC covers not only the nonresponse characteristics used
in the earlier USNRC TRCs, but also the presence of other data points that do not
fit within a standard probability distribution. Review of these data point outliers can
indicate the effects of context on the responses of crews. The TRCs are useful to see
the effects of context upon the TRCs; for example, the chapter contains TRCs from
the same plant modified by human factor changes to control boards. Also, changes
in a procedure for a given accident can lead to vastly different TRCs. TRCs can
be used as a tool to examine the effect of improvements or otherwise on crew performance. The context affects both the shape of the TRCs and deviations from the
random distribution curves, so it is possible to modify, by changes to MMI, training
or procedures to change the TRC shape, and this may also enhance the reliability of
the process.

By collecting simulator data, one can see the presence of both slips and mistakes. It is known that slips are easier to recover from and that it is more difficult to
recover from mistakes. Later in the chapter, there is a reference to some experiments
involving an emergency operating procedures tool. The results of the experiments
indicated that it helped to supply cognitive feedback to the crew and reduce the effect

of cognitive lockup. The tool supplies the feedback necessary for the crew to recover from mistakes, and this has been confirmed.

Because context is an important effect, the simulator data can be used to determine the influence of various context elements. The problem is the fixed structure of the training simulator. The simulator is primarily designed to support operator training, not for HRA investigations. However, over time it can be anticipated that there will be changes to procedures and to the control and instrumentation displays. Provided that data are collected over time based on careful observations of crew performance, data can yield insights into the effects of context. One HRA tool developed based on context is the holistic decision tree method (HDT) (Spurgin and Frank, 2004a) and was discussed in an earlier chapter in the book. The use of such models can incorporate context from observations or by the use of expert judgment.

In accident scenarios where crews' error rates are high, the HEPs can be used to check relationships between HEPs and context. Of course, it would be very useful to be able to carry out experiments in which context were changed in a controlled manner. Training simulators are really production machines and are not easily modified for the purpose of specific investigations. Modifications carried out by utilities are often to specific problems determined because of plant-related issues. Consider, for example, issues identified with elements of the EOPs. In these cases, changes are made to the EOPs and often the efficacy of the changes is not noted. Often the change is because of a technical issue with the EOPs.

The test of the emergency operating procedure tracking system (EOPTS) was one such case where it was possible to see the impact of an improvement in operator performance and reliability related to a support system. Here data showed the differences between the EOPTS and the paper EOP approach, as far as slips and mistakes generated by the crews were concerned. In this project, the crews were divided into two groups with one group using the paper procedures and the other group using the EOPTS. Time data and some observational data were collected. The errors made by the crews and their characteristics were noted.

Human reliability assessment (HRA) methods are essentially divided between generic models and plant-specific models, as mentioned in earlier chapters. Generic models are based on access to a given generic database, and the HRA user decides how to use the data in representing a particular human action (HA) or human interaction (HI) in the accident sequence. The analyst can choose to modify the given human error reliability probability (HEP) by the use of performance shaping factors (PSFs). The generic HEPs are justified by empirical and historical data and expert opinion is used to compensate for lack of appropriate empirical and HA context. The plant-specific HRA models rest on the use of expert judgment or insights derived from power plant simulators, which closely represent actual operator responses when faced with accident scenarios.

This chapter discusses simulator data collection methods used by various organizations. Later there is a discussion of how the simulated related HRA data/information have been used in various PRA/PSAs. In the latter part of the chapter, there is a discussion of the insights extracted from simulator studies and their relationship to

HRA. The insights cover more than just data, maybe the principle insight is related more to how operators respond to accidents.

12.1   DESIGN OF SIMULATORS

It is appropriate to briefly discuss what a plant simulator is, its design, its construction, and its use. Even in the early days of power plant design, mathematical models were constructed to study the safety of nuclear power plants by building models of both nuclear kinetics and the thermal behavior of the reactor core. Over time, the sophistication of these models has improved. About the same time, there was a need to design the control and protection systems for these reactor systems. In response to these requirements, which differed from those of safety analysis, there was a development of complete power plants simulations including modeling steam generators, pumps, and valves.

In designing the control systems initially, linear analysis methods were used based upon Bode and Nyquist methods. Later, it was considered that some nonlinear elements needed to be considered, so the move was made to simulation techniques. The early power plant simulations for this purpose made use of analog computers and computers came later. It was advantageous in that the trend to simulation of the complete plant for control and protection design purposes was useful in producing mathematical models that could be used for training simulators. For example, the Westinghouse Zion simulator (circa 1970) included a fairly comprehensive reactor core model and obtained experience from a person engaged in control systems design to help with the rest of the plant simulation models.

As time goes on, there is still a need to improve the reactor core models and accident-related models, such as modeling the complex steam/water break flow following a loss of coolant accident (LOCA). In the early days, say prior to 1960s, the emphasis was very much concerned with the dynamics of the reactor and even the importance of safety of the decay heat removal was underestimated, so much so that systems defined as safety related were those directly related to reactor heat removal and inventory control. The influence of the steam generators and auxiliary feed water systems on the safety of the power plant was underestimated, and these systems were not classified as safety systems. The relative importance of the reactor led to fairly sophisticated mathematical models of the core, but the remaining elements of the plant were significantly less sophisticated, such as having a table lookup to represent feed flow rather than a dynamic model of the pump, its steam turbine drive, feed water control valve, and associated control system.

An essential requirement for a training simulator is that it have the same look and feel as the real power plant control room. It is felt that to train the crews on essentially a different power plant may prevent the crews from functioning correctly. They would be in the position of asking are the indications the same as in the real plant? Not having a simulator producing correct plant indications could lead to uncertainty in the crew's decision-making when responding to plant indications during an accident. Hence there could be an increase in the risk in plant operations due to this feature..

The difference is that the power plant is represented in the simulator computer by a set of mathematical modes simulating the behavior of the plant during both normal operations and during accidents. In the simulator, the duplicate set of control room displays is driven by computers, however, the physical arrangement of the displays should correspond to the actual arrangements in the real main control room (MCR). The simulator designers follow human factors guidelines as far as trying to match the environment of the real MCR, so lighting, sound proofing and carpet should be close, if not identical to the real MCR. The appearance and operation of the simulator should duplicate the actual NPP main control room as far as the crews are concerned. A typical nuclear power plant simulator is shown in Figure 12.1.

The computer models representing the power plant should also duplicate the essence of the real transient behavior of the plant both in accident and normal operations. Also, the dynamics of the pseudo-displays should have similar dynamics to the real instruments. Things like data-logging computers and plotting printers should also be the same. When the operators go into the simulated MCR, it should have the look and feel of the actual plant. Sometimes the balance between the various components, in terms of how realistic each is, may be questioned.

In the early simulators, the balance between the modeling of the reactor and its systems and the balance of plant (BOP) was not correct. The reason for this was partially due to history and partially due to the unavailability of computers that could simulate the whole plant. The balance was toward modeling the reactor, which reflected the opinion of people that the reactor was central to the safety of the plant so the simulator design reflected their safety concerns. Over the years, the balance has been changing to be more reflective of the actual safety issues involving the rest of the plant. The early view of reactor and plant safety was reflected in the design basis accident approach to plant safety; later the use of PRA techniques led to a better appreciation of the interactions between systems and the role of humans in the safety of NPPs. This was further highlighted by the Three Mile Island (TMI) Unit 2 accident; see earlier chapter on NPP accidents.

FIGURE 12.1   A nuclear power plant simulator. (Courtesy of Pacific Gas and Electric. Used with permission.)

Another issue—is there a need to present to the operators all of the fine structure that safety analysts might need, because the details are filtered by the power plant and by the transfer functions of the instruments? The view of the plant as seen by the operators is different than that of the safety analysts. The view taken by the analysts is needed to examine all of the issues associated with the prediction of transient fuel and clad temperatures, to understand the impact of a given accident upon the NPP. Questions arise such as: will the clad be damaged? Do the automatic controls minimize or mitigate the accident effects? How do the actions of the crews influence the transient. The crews see the NPP behavior through the eyes of the instrumentation, and the fine structure of the thermal response, for example, is not seen. The crew needs only enough information to be able to decide what actions to take to terminate or mitigate the consequences of the accident. They do not need to know what percentage of the fluid leaving the reactor vessel is steam and how that percentage is changing every second. However, one should make the point that the closer the model is to the plant means that the simulator can be used for other investigations. For example, testing a new control system could make use of the simulator. What is seen as unnecessary modeling accuracy for operators becomes very important for control system tuning. For the tuning operation, the amplitude and phase of the feedback signals are extremely important to arrive at accurate controller settings.

12.2 SIMULATOR HISTORY

We forget that in the beginning of nuclear power, there were no simulators beyond investigators using plant simulators for accident studies and control system design investigations; for example, the Commissary Energy Atomic (CEA) of France built the first power plant model for use on an analog computer, and the U.S. Navy used a model of a ship-borne power plant for studies into the effect of steam catapult launches on boiler pressure. The first digital model of a gas-cooled reactor station was developed in the early 1960s (Spurgin and Carstairs, 1967). Some of these early analog simulations were used to examine the impact of induced faults on the capability of operators to control the disturbances.

Later both General Electric (in Morristown) and Westinghouse (at the Zion NPP site) built limited scope training simulators for the purpose of training NPP main control room (MCR) crews to be able to handle relatively simple accident scenarios using the event-based procedures. Later a few more simulators were added to the set of training simulators, but the numbers were limited and utilities often had to send their crews to training centers that now would be considered inadequate in that the facilities were based upon significantly different plants. Examples of this were control room crews from Connecticut Yankee NPP and Beznau being trained at one of the units at Zion Training Center. The control board arrangements were different, the plants were different, and the procedures were different. In the case of the Swiss plant, the displays were different, but at least the crews had their own instructors.

However, following the TMI Unit 2 accident, all was to change. The lessons learned from this accident led to a complete change in the attitude of the USNRC and the utilities. Before this accident, it was assumed that the automatic control and protection systems would take care of accidents and the operators were somewhat supernumerary. After the accident, the role and importance of the crews changed. In fact, the big message from TMI Unit 2 was that the crews and their environment were extremely important. Changes were called for to lead to the redesign of the MCRs, changes were made to the emergency procedures (from event-based to symptom-based), training of the crews, and the installation of plant-specific simulators to serve the station (provided the NPPs were the same) or to have specific simulators for given plants. Another evolution occurred in that scenarios used for training and evaluation were increased in complexity. The scenarios went from simple scenarios covering one or two failures to multiple failures. These later scenarios were difficult for the crews and were much more like real accidents.

Effectively, the majority of changes following the accident at TMI Unit 2 were human factors or were human reliability related. There were some equipment changes, but very few. The formation of the Institute of Nuclear Operations (INPO) occurred after the incident at TMI Unit 2, and this reinforced the idea that changes were needed in the human area. INPO caused attention to maintenance, training processes, and the function of management (organizational factors). The USNRC promulgated rules and regulations to take care of the other human factors issues.

12.3   BRIEF HISTORY OF SIMULATOR DATA COLLECTION STUDIES

Simulator data collection started with a USNRC-funded project, Safety-Related Operator Actions (SROA) project (Kozinsky et al., 1983) aimed at trying to set up rules for which safety-related actions could be taken by operators responding to accidents and which should be assigned to automation. The set of reports was fundamental in changing the course of early HRA studies, and introduced the concept of TRCs. A modified version of the TRC was incorporated into the Swain HRA handbook (Swain and Guttman, 1983). It also was part of the considerations that led to the EPRI human cognitive reliability (HCR) model (Hannaman, Spurgin, and Lukic, 1984b) and was part of the basis for the work of Dougherty and Fragola (1988). The data collection method used for the SROA project was the collection of the times at which operators took given actions from the start of the accident. General Physics developed a digital program to collect the times that actions were taken by operators during the operation of the simulator, and these records were changes of state of various switches. All successful response data for all crews for a specific HI are plotted against time. This is called the nonresponse probability curve:

$$P \text{ (nonresponse)} = 1\text{-}i/(N + 1) \tag{12.1}$$

where $i = i$th ordered point of the N data points in the HA set, omitting nonsuccessful crew responses; P is the probability of operator nonresponse.

The accent in these early simulator studies was on data collection based on time responses of the crews to accident scenarios. It was believed at the time that TRCs represented the cognitive responses of crews, and the extension of the TRC could be used to predict the cognitive error at the end of a time window. A TRC can be considered to represent the random variation in the action times of persons performing a given task. This type of curve holds for humans performing tasks as widely ranging as programming computer programs to participating in running track events.

In a similar time frame, EDF was carrying out simulator studies (Villemeur et al., 1986), but with the emphasis on observations of crew actions, for example, to determine the acceptability of a reactor engineer support system, rather than concentrating on time responses. EDF's approach was to study several different scenarios using only a small number (two or three) of crews. The EPRI approach was to use all or most of a station's crews and a small number of scenarios (three or five) during the ORE program. However, the scenarios were complex and consisted of two or three parts of increasing complexity.

EPRI funded a number of simulator studies following the development of the HCR model. The principal one was the OREs (Spurgin et al., 1990a), and this was aimed at the verification of the basis of human cognitive reliability (HCR) correlation or model. This study was followed by a similar study called the Operator Reliability Experiments for Maanshan NPP (Orvis and Spurgin, 1990) undertaken jointly with the Taiwan Power Company (TPC) to help TPC with improvements in training and insights into operators' actions to help TPC in the area of enhanced safety of their plants. The Maanshan NPPs are three-loop NPPs designed by Westinghouse Electric Company. Another EPRI/TPC program was a study (Spurgin et al., 1990b) into the effectiveness of an operator support system for BWR EOP called the BWR EOPTS at the Kuosheng (Taiwan) NPP. The Kuosheng NPP is a boiling water reactor NPP designed by General Electric (GE).

Following these studies, a number of organizations repeated simulator data collection exercises based on the techniques used in either the SROA or ORE studies; see IAEA report (Spurgin, 1994).

Although there were developments in the ORE project related to data collection, the main motivation was to verify the basis of the HCR model. A number of simulator studies have confirmed the limitations of the HCR model and issues with TRCs in general. However, a great deal of insight into operator performance can be gathered based on the TRC data. The TRC is a useful tool for instructors and training managers to focus their attention on the competence of the crews and the impact of training, procedure, and MMI deficiencies. Comparative analysis of results from similar plants can point to design improvements to improve crew responses to the same accident type.

Studies have been carried out by Hungarian (Bareith et al., 1999), Czech (Holy, 2000), and Bulgarian (Petkov et al., 2004) investigators with a view to help understand the performance of control room crews and the influence of accident scenario effects on their reliability to accurately perform necessary actions to mitigate or terminate the effects of an accident. In addition, some studies have included modification of EOPs from event based to symptom based. This draws on the insights gained from these studies and others. The Chinese focus has been twofold: improvement to TRCs for a Chinese reactor PSA/HRA study, and an examination of the bases of the HCR model. The Czechs were interested in the design and improvement of the symptom-based EOPs versus the earlier event-based EOPs. The Hungarians devoted considerable effort to collecting data at the Paks power plant simulator for several reasons, but mainly to support their PSA/HRA. They examined the reliability of crew actions over a number of separate simulator sessions and included sessions involving both event- and symptom-based EOPs and the move to digital-based safety protection instrumentation. The Bulgarian studies focused on the impact of individual operator context and crew-based operations upon reliability.

There are some limitations associated with the use of power plant simulators from a couple points of view: one is associated with collection of human error rates, and the other is in the research into the impact of changes in training, MMI, or human–system interface (HSI) procedures on human performance. This is because of the characteristics of simulator training centers. There are few opportunities for changes to training methods, procedures, or man–machine interfaces. However, if changes occur, then at these times one should take the opportunity to carry out comparative studies of before-and-after effects. Using simulators for human error rate purposes is not very convincing because the number of crews and events are rather low to predict HEPs of the order of 1.0E-01 or lower. Even for a plant with 36 crews (Zaporizhzehya NPP in the Ukraine) with few failures, the minimum HEP is 1/36 (~0.03) when a single failure occurs. Of course, if there are a large number of failures (13 crews out of 15 crews failed), such as occurred on one occasion, then the estimated value is 13/15 = 0.87. In this case, the utility took steps to reevaluate the procedure and the training of the crews. The solution was to do the following: clarify an instruction in the procedure, add a logic diagram to the support materials in the procedure, and enhance training in understanding the basic logic of the operation and the plant requirements. Following the changes, there were no failures in the accident scenario during tests carried out much later.

The point of the Czech simulator tests was to check the improvements in going from event-based to symptom-based EOPs. The Czechs were in the business of continual improvements to the EOPs from both technical and human factors (HFs) points of view. Data collected during simulated accidents were used to evaluate both technical and HF deficiencies. Incorrect actions can result from technical errors in the procedures, and can also result from problems because of the layout of the EOPs. The descriptions of the EOP steps can be incorrectly interpreted by the crews and lead to human errors, just as much as technically incorrect steps can lead to errors.

Likewise, the Hungarians investigated changes in going from event-based to symptom-based procedures as well as the effects of improving the control and protection equipment in moving from analog to digital-based equipment. Studies were carried out at the Paks NPP simulator center. The early tests were based on the EPRI ORE methods together with a Paks-designed data collection method integrated into their simulator for use by their instructors. Other simulator tests followed over the years, along with improvements in data collection including the use of bar code methods for collecting information based on instructors' observations of crew responses (Spurgin, Bareith, and Moeini, 1996); the project was paid for by the U.S. Department of Energy.

Such opportunities have occurred in the past, and we will try to indicate the impact of such changes on crew performance. Sometimes, two similar types of plants do not have exactly the same methods for dealing with accidents, so we can examine the responses of crews to highlight the ways the crews respond differently. The next interest is in how these bits and pieces can be integrated in a coherent manner to advance the HRA process.

12.4   PURPOSES OF DATA COLLECTION

The main purpose of MCR simulators is for the training of control room personnel, but they can be used for a variety of other purposes. Evaluation of the uses depends on what data can be collected for analyses. The range and quality of the data available depend on the type and extent of the data systems connected to the simulator and associated with its software. Typically, data are collected on the plant variables and some of the operator actions with time as the independent variable. The newer simulators have more state-of-the-art recording capability comparable to the actual power plant. The simulator facility makes it possible to collect data concerning all of the MCR operator and instructor actions and movements, conversations, and sufficient analog and digital data. Here analog means continuously varying plant variables, and digital means discrete changes in state relating to operator actions, pumps turning off or on, or faults (steam generator tube ruptures) being generated by the scenario-designed control program. The scenario control program is usually designed by the instructors to ensure that the scenario is consistently presented to the various crews. The program selects the initial conditions, plant state including power level, and availability of safety and other equipment. The series of faults to be injected and their timings are set in this program. All aspects of the scenario are tested before the tests. Often the control program is "canned" for future use to minimize repeating the work.

Data requirements are as follows:

1. For the purpose of judging operator performance and crew performance in responding to a variety of accidents
2. For estimates of crew performance for HRA purposes (additional efforts should be made in order to overcome differences between data requirements for HRA and training purposes)
3. For comparison purposes between plant transients following accidents or incidents and simulator transients; this is to yield better insights into the progression of the real accident or incident
4. The accidents or incidents can be used for simulator validation purposes (fidelity requirements stemming from USNRC rules); as part of this requirement, users must show that the simulator reacts the same over time, and there are only documented changes to the simulator programs
5. For verification and validation of operator performance as a part of licensing requirements
6. For verification and validation of EOPs
7. For checking human factor improvements or changes

8. For performance evaluation of display system usage by crews following both display equipment and software changes, where it can affect response timing to plant or operator inputs

12.5   DATA COLLECTION SYSTEMS AND PROCESSES

As mentioned above, the data are of two types: simulator related and observer related. In addition, video recordings are usually taken to provide backup for the observer data. If the instructors or observers wish to check some of their observations, the videos can be rerun. In addition to the collection of data related to the simulator exercises, information is also derived from "postmortem" interviews at the end of the simulator sessions. These postmortems yield insights into crew perceptions about the accident scenario and possible causes for their errors. Often the videos or parts of the videos are rerun during the postmortems either for clarification by the instructors or for discussions by the crews of issues that come up during running the accident scenario. A bar code reader can be used to identify video times relative to operator actions. The SROA project (Kozinsky et al., 1983) collected simulator data on tape units. The data were processed offline to produce the TRCs.

12.5.1   MANUAL DATA COLLECTION

The ORE project, like many subsequent projects, captured the time by stopwatches at which certain operator actions were taken. For these projects, manual data collection was used. Later, at one PWR plant, response time was collected using a data collection program that was part of the simulator computer package supplied by the simulator designer. Following the first simulator sessions, one of the "lessons learned" was the need to collect observational data as well as the action times. The process was to analyze an accident and determine the key operator actions and their sequence. For each key action, the associated alarms or indications were identified, and a listing was made for the observers to follow and help them identify when the key operator actions were taken. Space was allocated for the observers to make notes related to the actions taken by the crew.

12.5.2   AUTOMATED DATA COLLECTION SYSTEMS

Insights gained from the analysis of simulator accident sessions during the ORE project were of particular interest to a training manager at a West Coast PWR power plant. He wanted to get the process automated, because the overhead burden was too labor-intensive for a training establishment; however, he was very interested in capturing the same kind of insights that were obtained during the data collected as part of the ORE program. The development of an automated system to collect time and observational data was supported by EPRI and the utility. The digital program went through a number of development steps to meet the needs of the manager and those of the training department. The main requirement was to design an effective collection system that encompassed training department requirements with regard to matching INPO Academy specifications.

Of the digital programs developed, the first version was called OPERAS (Spurgin and Spurgin, 1994). This was a joint project funded by EPRI and funded and supported by Pacific Gas and Electric. The program collected both analog and digital data. The analog data were made up of simulated plant variables, such as reactor pressure and temperature. The digital data were discrete changes in state, such as switch position changes. The program was based on the Microsoft Excel® program. Excel was used because of the flexibility of coding, while changes were being made to match the developing needs of the power plant. The designers worked closely with training department personnel to evolve a useful system for principally training use; however, it could also be used for HRA purposes. Later several improvements were made including reprogramming in C++ code to increase speed and reliability, the introduction of bar-coding for the capture of observational data, as well as several changes to the kind of observational data to be collected, resulting in the CREDIT data collection system (see Figure 12.2). The information displayed was also modified to the needs of the instructors. Time response data and the corresponding operator actions were also output. This data along with the video records were found to be useful in helping the instructors go over issues related to procedure use and the prioritizing of operator uses of display information. Having the data available also enabled the operators to go over in detail their view of what happened during the transient.

Over a period of time that these data systems were being developed, the style of the instructors changed from being concerned with pointing out errors to where they and the shift foremen led the discussion as to the sequence of events and the crew's responses. This process was much more useful, and the role of the instructor was changed to where the instructor was there to help the operators achieve the best performance to operate the plant both safely and economically. It is interesting to point out that U.S. plants are now operating at 90+% availability compared with what they were accomplishing (60% availability) in the 1980s.

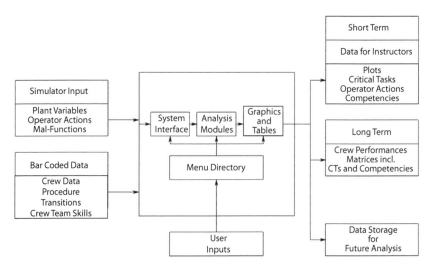

FIGURE 12.2   Automated data collection system (CREDIT).

The final version of the data collection system is called CREDIT Version-3.1 (Spurgin and Spurgin, 2000). For PG&E, the observational data collected were based on the INPO specifications for crew training needs.

The CREDIT data collection system combined with a bar code observation data collection system can capture time data, plant response data, and observational data associated with procedure use, problem areas with the display and control systems, and information associated with crew communications and their skill in observing plant alarms and indications. The system can compare an expected sequence of operator actions responding to an accident with the actual responses taken by the crews. This is very useful information for instructors to present to the crews during the postmortem after the training exercises. The user can indicate by bar code times of interest, so one can go to the video recordings and find the area of interest very quickly and show it to the specific crew for their comments. All of this information is of interest to the training group, including the training and operations managers as well as the HRA specialists.

The Paks NPP simulator personnel also had a simulator data collection system used by instructors for training purposes. It consisted of time plots and operator action times. It was quite an advanced system at the time of the first Paks NPP simulator experiments (1992) and supported the work done to collect times for operator actions in response to simulated accidents (Spurgin, Bareith, and Moeini, 1996). Observational data were recorded manually initially; later a bar code system was developed to collect observational data (Spurgin, Bareith, and Moeini, 1996).

The move to collect observational data was to get some insight into the reasons for crew deviations and errors as seen during some crew responses to accident scenarios. Also included was the move to second-generation HRA methods, which emphasized context and increased interest in trying to see the effect of context on operator responses. Consequently, any data collection system must account for "how the context affects actions on a second-by-second basis" (Hollnagel, 1998). The need for detailed and dynamic determination of operator's performance context leads to the need for continuous monitoring, diagnostics, and auto data mining in control rooms or simulators, and helps to reduce the use of experts to make judgments on the effects of context (Spurgin and Petkov, 2005).

12.6   INSIGHTS: FROM TIME AND OBSERVATIONAL DATA

In the first series of simulator data collections (Kozinsky et al., 1983), the investigators considered only the successful operator responses and used this data to construct the nonresponse curves. The number of successful actions (N) was used to determine the response probability; see Equation 12.1. Of course, some actions were very successful (i.e., all the crews were successful); in other cases, some crews were unsuccessful up to the time that the simulator session was ended. The original study, funded by the USNRC, was interested in whether or not operators were reasonably successful within a time window; therefore, it appeared to the investigators that many actions could be left to the operators. They provided a useful backup to automated systems for many safety-related actions.

However, with the increased importance of PRA/HRAs, the use of simulators for HRA studies gained in importance. The examination in detail of the simulator data was pursued, and it was realized that the outlier data and unsuccessful data were worth examining. The outliers from the point of view of what was the cause of why they did not conform to the random response curve, and for the failures, what were the number and causes. In fact, it was the study of the outliers that led to the development of the HDT approach and also the realization that the TRC was not a measure of cognitive performance but rather the variation in the performance of operators. The ordering of the crews in response to accidents was itself variable. In other words, the "fast" crews were not always first in responding. Some crews were better in some cases, but not always. It is interesting to note that data relating to track performance (i.e., 10k performance) shows the same kind of variation, and the distributions can be broken into three or four classes, such as professional, strong amateurs, amateurs, and not-serious. The author's interest in looking at the data was because he used to run 5ks, 10ks, and marathons. The analysis of this data was not too rigorous, however.

Figure 12.3 shows a typical TRC. The figure shows both the nonresponse curve and the full success curve, including the presence of two outliers. Of course, when running a simulator, one is limited as to how long one would wait for a crew, which was having difficulties, to be successful. In these circumstances, the training session would be stopped after about 1 hour and a half and the crew judged to have failed to control/terminate the accident. In an actual accident, the crew may recover after some time, but the consequences might not be good (TMI Unit 2). One can immediately appreciate that extrapolation of the nonresponse curve to intersect with a time window of about 300 seconds would yield a very low probability value. Yet the use of the actual curve, including the outliers, does not indicate the potential for the same probability number. For example, if one crew failed out of 24 crews, the

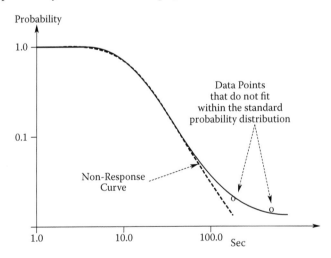

FIGURE 12.3 Typical time reliability curve.

probability would be 1/24 = 0.04. This is vastly different than 1.0E-03 indicated by the intersection of the nonresponse curve with the time window.

Of course, examination of the time data tells us nothing about what has influenced these two crews, so that their responses do not fall within the "normal" random response curve. This is where observations can supply some insights into what appears to be additional delay to expected response times. One can expect some variability in data because of small data sets. There are some six crews/unit, so there are only six data points for a power plant with only one unit. One of the NPPs with the largest number of crews is the Zaporizhzehya NPP Ukraine, with 6 units and 36 crews. The Paks NPP has 24 crews. So with this number of crews, one can draw quite a good curve.

The collection of observational data now becomes important to understanding both the normal and abnormal responses. Also, the move to context-based HRA methods calls for information about how the crews react to the HSI, procedure layout and details, training, intercrew communications, and experience of accident scenarios. The details of the responses are informative for training personnel and also can fulfill the needs of the HRA experts. Some discipline is required on the part of the observers to generate observational data that is useful and balanced between observers. The use of a taxonomy helps to generate a useful database capable of being input into a computer. Once the database has been assembled, the user can extract data that can be used for modification of training processes and can be used for objective measures for context influences.

TRC data plots can quickly expose the impact of changes to the design of the HSI and its procedures and training. Changes can occur both to the TRC and to the distribution of outliers. Figure 12.4 shows the effects of both a human factor change and a difference in the way a particular operation is carried out at two different PWR

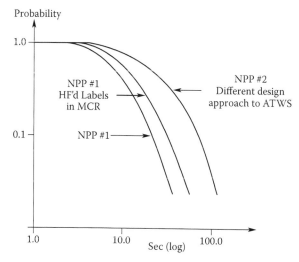

FIGURE 12.4 A set of anticipated transient without scram time reliability curves for different plant conditions.

power plants. The particular examples are for the response to an anticipated transient without scram (ATWS). One curve is the basic response of the crew to an ATWS at a U.S. PWR plant. The response of the crew is rapid. The actions taken by the crew are to observe that an accident has occurred (turbine trip) and the controls have not been tripped and dropped into the core. The actions of the crew are to try to trip the reactor by pressing the scram buttons, then observing that if that did not work, they would achieve the same effect by tripping the breakers of the power supplies to a number of components, including the control rod drive power supply. The reactor control rods then dropped into the core and shut the reactor down. (Note that this was not a case of rods jamming and therefore not dropping.) The response was well learned and there were no delayed responses beyond the normal random variation. In the case of using the power supply breakers, once the rods have dropped, the crew would reclose the breakers so that pumps on the same bus could operate.

The variation in the TRCs associated with #1 was due to a change in the labels as part of a human factors improvement to the control boards. The change was from one designation of the breakers to one more closely associated with a numbering system given on drawings. The result of the change was an increase in the execution time of the crews. All crews were affected, but the distribution of the crew times is different, in that some crews were more affected by the change than others. It is interesting to note that the responses were still very rapid in both cases, as a result of training emphasis and design of the details of the actions to be taken. The crew responses were more consistent for the first session, but there was more uncertainty in the second set, because of hesitation related to the labels. It is expected that the crew responses will be nearly identical to the first set of results after a short time relearning the label system. At the time of the session, the crews were a little uncertain because of the labels. An estimate of the HEP for this operation would be 1.0E-03 or less. Compare this HEP with the HEP derived from Swain's TRC for a time window of 1 minute, in this case the HEP estimate would be 1.0 (i.e., the crews would always fail), whereas the results show exactly the opposite. The operator actions are both quick and accurate, due to training and the clear arrangement of controls on the boards, with a little hesitance due to the labeling change. But even in this case, the operators went to the correct breaker switches but had to check that the switches were the right ones before taking action.

The NPP #2 curve shown in Figure 12.4 relates to the responses of the PWR Taiwan plant crews to an ATWS. The Taiwan plant had a different philosophy or design for dealing with an ATWS accident. To respond to the ATWS, the crews are expected to remove power from the control rod drive mechanisms. If the reactor trip breaker does not open, then the operators are expected to remove power from the drive itself. However, in the case of the Taiwan PWR, the breakers for the drives are in the auxiliary control room. It takes some time for an operator to go from the main control room to the auxiliary control room and open the breakers. So, although the steps in the process are the same as they are for the U.S. PWR plant, the need to go to the auxiliary control room meant that more time was taken by the Taiwan plant crews to carry out this action. For the ATWS accident simulator sessions, the role of the operator going to the auxiliary control room was played by the instructor, and he waited for a time consistent with the actual time that would be taken by a crew member before he opened the simulated breakers. Since the training, procedures, and instrumentation are very similar to the U.S. plant, the estimated HEP would be similar.

The ATWS accident has received a lot of attention in the training programs. The USNRC has directed that the crews be trained to recognize the accident quickly and take action. Crew training on the ATWS accident has focused upon both classroom training and simulator sessions. It appears that the operator responses to the ATWS accident are both fast and reliable. In part, this is due to the emphasis placed upon this accident by the USNRC, but the indications and actions required to terminate or mitigate are very clear, and the operators are well trained in the use of emergency procedures.

The steps in the process are as follows. The turbine trips, and as a consequence of this the reactor should be tripped and the control rods drop into the core, supposedly within 1.6 seconds. However, the trip breaker does not open and the rods do not drop, so the reactor stills generates power and the safety valves on the primary and secondary sides open. The crews recognize the request for reactor trip, an important parameter. The crews then check that the rods position indicators have not dropped. Seeing the indicator position is not correct, they take steps to open the breakers that supply power to the rod drive mechanisms. It is here that there were differences between the U.S. PWR and the Taiwan PWR.

Here we are interested in understanding these actions and the reasons for operators' high reliability for this transient. The reasons are a clear indication of a need (reactor scram window), a failure to insert control rods (reactor control rod display), and a clearly defined path to achieve success (trip power supply for rod drives). This sequence of events is often practiced on the simulator. For a specific PWR, this was accomplished by all of the crews within about 25 seconds.

In the process of carrying out simulator sessions, a number of issues associated with EOPs occurred from minor training-related issues to major problems with converting from event-based procedures to symptom-based EOPs. The old event-based EOPs required one to diagnose an accident before one could select the appropriate procedure to deal with the accident. If the accident could not be diagnosed, then it was impossible for the crews to respond. The diagnosis of an accident was left to the crews rather than having a diagnosis procedure, so it was difficult for the crews. The idea of having a separate procedure for each accident was a constraint. So the approach was to identify a range of loss of coolant accidents (LOCAs) and have a procedure for each, so the process became clumsy and did not address the problems associated with complex accident scenarios. These issues were addressed in the move to symptom-based procedures that came about after the TMI Unit 2 accident.

In general, the symptom-based EOPs contain a better diagnostic routine and a way of dealing with multiple failure accidents. In moving from event-based to symptom-based EOPs, the root cause of errors has changed from diagnostic to interpretative errors. No one can see even cases where a crew knows that a steam generator tube rupture (STGR) accident has occurred, but the procedure reader has trouble seeing the transition from the diagnosis part to the STGR response part of the procedure.

In classical discussions on human error, there is always talk about "slips" and "mistakes." Slips are due to the crew having the correct intention about an accident but a problem arises from failure to perform the action correctly. A mistake arises because of an incorrect understanding about the accident and the wrong action is taken. One problem with using simulators is how to evoke a situation, which can lead to a slip or mistake. They can be observed and evoked by the use of complex accidents. The thing that has been observed is not totally clear on just how to ensure that a slip or mistake can be forced. Even asking the instructors to suggest which scenarios will lead to a slip or mistake can be unsuccessful.

The capability of the scenario designers to craft scenarios that can lead to slips or mistakes is limited. One can design certain scenarios, and it is likely that slips and mistakes could occur, but training of the crews can be such that they do not occur. Often even for the instructors the crews performance is not entirely predictable. However, the psychology of slips and mistakes has been pointed out by various researchers in that slips are easier for the crews to recover from, and this is confirmed by the simulator studies.

In an investigation funded by EPRI and Taiwan Power Company (TPC), a comparative study was made with the use of symptom-based procedures with the crews using them in either a chart form or based upon a computer using the EOPTS (Spurgin, Orvis et al., 1990b) approach. The EOPTS was an artificial intelligence approach based upon plant data derived from the simulator. The data were plant variables, analog variables (pressure, temperature, and flows) together with digital variables (switch positions, etc.). The application to a real plant would gain similar data from plant sensors and various digital outputs like switch contacts.

The results of the study indicated that the EOPTS was much better at helping the operators recover from mistakes. In general, EOPs have recovery paths, and crews are instructed to return to the diagnostic routine if the accident is not being dealt with successfully. However, in practice, it is difficult to get crews to return to the beginning. In addition, the characteristics of the accident can change with time, making use of the procedures more difficult. The image that the crew has formed limits their ability to recover. In the case of slips, recovery is much more likely, because the operators have a correct image of what the plant state should be, and they can compare it with the actual state, seeing different acts as an incentive to reexamine their actions. In the case of mistakes, the crews' mental image of the situation is incorrect. Table 12.1 shows the results of the EOPTS study for 28 crew/scenarios.

The EOPTS crews are twice as good as the flow-chart crews in recovering from errors, despite the fact that the EOPTS was in a development stage. But the data also show the difficulty of operators using paper EOPs to recover from mistakes. An examination of the errors divided the group into slips and mistakes. From the

TABLE 12.1
Errors and Unrecovered Errors

Item	Flow Chart	EOPTS
Total errors	23	11
Unrecovered errors	15	3

observations, the slips were fairly easily recognized by the crews and recovered. The number of slips made by crews using the charts was about the same as those using the EOPTS, and the recovery rate was about the same. It was only for mistakes that the EOPTS crews had a clear advantage.

12.7 USE OF SIMULATOR DATA FOR HRA PURPOSES

The simulator is very useful for some purposes, but using it in the same way that we gather equipment failure data to estimate equipment failure probabilities is not correct. An in-depth review into the uses of simulators and the mining of that data is contained in the references (Spurgin and Petkov, 2005). The crew failure rates depend on the context under which the crew operates. The context for crews encompasses different NPPs with different control room arrangements, different training methods, different procedures, different management, and different communications protocols. The environment for control crews can be vastly different from one plant to another. The human reliability data set for a given plant is limited because of the limited number of crews per station (i.e., six crews per unit).

In estimating the HEPs associated with a given accident scenario, one is trying to estimate the impact of the context under which the crews operate. The simulator gives a way to observe the crews' performance and to try to understand which aspects have the strongest effect on the crews. Examination of the observational data can help the HRA specialist determine for a specific accident scenario which fits the best description of each of the IFs. Table 12.2 lists possible definitions of the quality of one of the IFs. The list was used for a space project, but command, control, and decision making can be considered equal to the combined roles of the shift supervisor and procedure reader of an NPP. Observation of the crew and its relationship to the procedure reader and the shift supervisor during the training session can give insights into which quality factor best represents the activities of the crew.

For a given accident, the HRA analysts would consider the ranking of the IFs along with the corresponding QFs. Measures of the relative importance of the IFs and which QF should be selected for a given accident are determined by review of the

TABLE 12.2
Description of Typical Quality Factors

Quality Factor	Efficient	Adequate	Deficient
Command, control, and decision making	Well-defined lines of authority, established protocol, and experienced leaders, team used to working together	Defined lines of authority, but some uncertainty, established protocol, but not such experienced leadership, some experienced personnel used to working together	No clear lines of authority, inexperienced team members or ambiguous protocols, inexperienced leadership

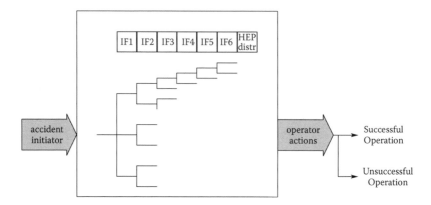

FIGURE 12.5   Structure of a holistic decision tree.

simulator records, particularly the observational data. For example, if the procedure incidents occur more than training issues, then it is likely the HRA analyst will bias the IFs toward the procedure IF rather than the training IF. If the procedure deviations have safety implications, then the HRA analyst is likely to consider that those procedures related to the accident are not very good. If the problems with the procedures stem from the same area for different scenarios, then the problems are more likely to be procedure related. Whereas if the same area is satisfactory in the case of one scenario and unsatisfactory for another scenario, then the problem may be with training. One can see the role that analysis of the simulator observations plays in determining the impact of context and, hence, the determination of the pathways through the HDT model. The structure of the HDT including both the influence factors and branches, which depend on the pathway through the HDT, are shown in Figure 12.5.

12.8   DESIGN OF SCENARIOS

An important feature of the ORE program was the design of scenarios to be used to collect data on the responses of operators to accidents. A particular myth in the HRA field was that the responses of operators to simulated accidents were not representative of how operators would respond during "real" accidents, and therefore simulators, providing insights into operator responses, were not credible. Another point made was that the operators expected to be exposed to large accidents and were ready; therefore their responses were not representative. It is agreed that there is some justification for these views based upon early simulator usage. It should be pointed out that if the simulators were not highly useful for preparing the crews, the industry was paying a lot for making simulators as close to the real thing as possible, and the cost of having crews take part in accident simulations would be wasted.

There are four points to be made in refuting the critics:

1. Crew training has become much more professional over the years, and exposure to simulated accidents is an important part of crew training. Earlier crews were inclined to think of the simulator as a computer game, but this has changed. The simulator has become a central tool in the preparation of the crews along with requalification for their jobs.
2. The design of simulator scenarios has changed from the big accident, such as the large loss of coolant accident (LLOCA), to something more subtle involving multiple failures. These types of scenarios are more realistic and representative of accidents.
3. Improvements in the simulator models and components have been seen by the crews more closely representing the plants; therefore what is presented is very close to plant responses and not just a game.
4. There have been circumstances in which simulated scenarios created to test the crews have actually occurred subsequently, and the crew responses were identical to what happened before in the simulator.

In the ORE program, opportunities for testing the crews were limited, so the test scenarios were designed to make the best use of each opportunity and present the crews with a number of different situations. Each test scenario was made up of two or three parts. The first part would be an instrument or controller failure, the second part would be a more severe accident but one from which the crew could be expected to recover from and be capable of returning the plant to its initial condition, and the third part would be a severe accident in which the requirement was to put the reactor into a safe state, such as hot shutdown. Previously, the latter scenario was thought of as being the only kind of test scenario.

By designing a multiple scenario made up of these three parts, information could be gained on how the crews responded to different parts of the scenarios. Some were quite easy and others were more difficult scenarios. The crews, as the scenarios developed, became more interested in responding and less conscious of nonplant personnel observing their actions.

The expected result of a failure of a controller or sensor was to place the controller from automatic operation to manual operation, and for the crew to manipulate the controls as required.

A minor accident, such as a pump or component failure, falls into the second group of accidents to be used in the design of the test scenarios. The crew is expected to be able to diagnose the accident and return the plant to its initial condition within a short time. The accident does test the cognitive powers of the crew to diagnose the problem and then take actions to start standby pumps or bypass failed components. The power plant goes through a transient; the crew recovers the plant and then returns the plant to the same load state. A pressurizer spray operation or an isolatable relief valve on the pressurizer is such a minor accident that can turn into a more severe accident if it is not dealt with, so it presents a hazard but can be recovered and the plant operated manually.

The last part of the complete scenario is the initiation of a plant failure that cannot be restored, like a steam line or steam generator tube rupture. Not only was there one of these types of accident scenarios, but other plant equipment states were set up, so, for example, the auxiliary feed pumps were out of service or a pump failed on demand. This type of transient was more difficult than the straight large accident. It was the operators' response to these types of scenarios that ORE was all about. In many of the cases, the operators cannot continue to operate the plant as before, so the plant has to be shut down and the break has to be repaired. These large accident scenarios are the kinds of scenarios that were previously run without the prior transients. The combination of the three parts gives a lot of data on how the crews operate and can be fitted into the normal scheduled time for what was called requalification training. The role of instructors was to run the simulators, collect data, and observe the actions of the crews, but not to assist the crews during these periods.

The ORE program ran three to five such complex scenarios per plant and used all of the available crews. Normally, there are six crews per unit, so a two-unit station would have 12 crews. In some scenarios, all responses were successful, and in others some crews would fail in the middle scenario and we could not proceed to the classical large accident. The scenarios had the purpose of presenting the crews with a range of scenarios that called for a variety of expected skilled behavior. Remember the ORE program was set up to attempt to test the validity of the HCR model, so the researchers were looking for situations that could be reasonably expected to correspond to skill-, rule-, and knowledge-based behavior. Part of the design of the scenarios was to prevent the automatic protection actions from terminating an accident, so various pieces of equipment were either taken out of service or failed upon being requested by the control and protection system. This meant that the operators were forced to use either the abnormal or emergency procedures to tackle the accidents.

It is interesting to note that there was no sign of the early crews passing information on to the later crews, as judged by their responses to the multiple accident formulation. The reasons for this lack of transfer might be that it was in the interest of the crews to bypass the impact of gaining information to honing their skills, the design of the complex scenario was not helpful to passing on details, and the crews did not want to support the following crews, because this would leave the plant less safe because some crews were not sufficiently good to ensure the safety of the plant.

The USNRC has been concerned with the crews passing on information that the stated crews being judged for reactor operator (RO) licenses had to be exposed to different scenarios. The advantage of comparing the operators' performance based on a series of accidents was lost. In the author's opinion, this is a loss to the plant and training management to be able to see if the training program is working effectively and crew performance is steadily improving. Also, because the opportunity for the crews to be trained on given scenarios is limited, the distribution of test scenarios could be examined to see if the SAT requirements were being met. In other words, the best scenarios could be reduced in frequency and the less-effective scenarios increased to level out the performance of the range of scenarios.

12.9   DISCUSSION AND CONCLUSIONS

A dynamic theory of context considers it as a dynamic fuzzy set of entities that influence human cognitive behavior on a particular occasion. The HRA methods of all generations try to model the contextual effects on human action (HA) by structuring (tree/graph) context and changing expert quality level judgments with fuzzy assigned factors. For example, the THERP (Swain and Guttman, 1983) method applies HA decomposition (HRA event tree), PSFs, and dependence adjusting; the HEART (Williams, 1988) method uses EPC, effect factors, and their guessed fractional proportions; the CREAM method expands the fuzzy approach by using CPCs that should be coupled with each of cognitive subprocesses (functions) and weighed CPCs for control modes and cognitive functions (Hollnagel, 1998); and the HDT method (Spurgin and Frank, 2004) models contextual effects by IFs and QFs.

Simulators are tools that can be used to elucidate a number of issues by the use of observations, measurement, and postsession discussions with operators. Unfortunately, NPP simulators are mostly fixed environments, so that the opportunity for modification of the environment is limited. However, more flexible simulators are available for testing purposes at Holden and in Japan, so these units might help in the advance of HRA technology. Currently, many of the modifiers used in the various HRA methods, such as PSFs, have to be estimated by expert judgment. Instructors can be used as a good source for estimates when it comes to certain estimates, such as impact of EOP changes. Other estimates are difficult for instructors to make because they are not exposed to variations in these particular changes. The Halden and MITI flexible simulators are not discussed in detail. Both machines have computer-generated displays that can be potentially modified to present different displays and arrangements and therefore could be used to examine the impact of HSI changes and also the impact of various computerized EOP approaches. In fact, Halden has such a computerized EOP system, but there needs to be close testing of various forms to see what benefits accrue to any particular implementation.

EDF redesigned a new operator support system for their N 4 reactors and tested it extensively, but as far as is known, no comparative performance evaluations were released. The new system was a major change from the board arrangements and paper procedures of the prior reactor systems, such as the 900 MW (e) three-loop PWRs. Most of the work was trying to get the new system working and checking its performance.

It would seem that some limited small-scale tests might be the way to go. The evaluation of complete changes to the HSI and the procedures leads to problems because of the required changes in training and also the need to evaluate the trained state of the crews for each of the anticipated changes. For our examination of the EOPTS (Spurgin et al., 1990b), the paper procedures were changed to GE column set, and EOPTS was programmed to represent those procedures. Training was provided for each of the crews on either one or the other of the approaches to be investigated. In a way, they were equally prepared in two new approaches. This is a little better than one group using well-known EOP formulations versus the other group having very little experience with the computer-driven approach. As it happens, the operators liked both improvements, but cognitively they did better with the EOPTS.

13 Impact of Systematic Data Collection on Training

INTRODUCTION

Most training organizations use the concept of a systematic approach to training (SAT). This means taking a systems approach with respect to the training process. The program consists of four elements (after Barnes et al., 2001):

Systematic analysis of the jobs to be performed:

- Learning objectives derived from the analysis, which describe the desired performance after training
- Training design and implementation based on the learning objectives
- Evaluation of trainee mastery of the objectives during training
- Evaluation and revision of the training based on the performance of trained personnel in the job setting

The last bullet on evaluation and revision is the most important aspect of the SAT approach in my mind in that it does two things: It covers the performance of the personnel and it provides a mechanism to improve the training program itself. Without this feedback mechanism, the training of the personnel may have no relevance to the real needs of the personnel to successfully combat accidents and ensure the safety of the plant.

The power plant simulator provides a great tool for the station management to evaluate the station personnel's performance in responding to accidents. However, the value of the simulator is diminished if records are not collected on the performances of the station crews in carrying out their duties during simulated accident scenarios.

In its attempt to ensure that U.S. Nuclear Power Plants have qualified staff to operate their plants, they have instituted an accreditation plan

In the United States, a nuclear power plant (NPP) licensee's program is USNRC approved when accredited by the National Nuclear Accrediting Board. There are currently nine license training programs subject to accreditation. In this section we are concerned with the control rooms and the relationship of the use of simulators, effective training of control room crews, and the impact of the HRA estimates. Here, only those persons forming control room crews are considered:

- Licensed and nonlicensed operators

- Shift supervisors (SSs)
- Shift technical advisors (STAs)

The training of control room crews is divided into two groups: the initial training and continuing training groups. The main focus here is continuing training and how to improve the performance and reliability of the process by the analysis of data collected during training sessions. Specifically, the data collection and analysis process will be covered along with how it can be used to enhance the performance of the individuals, crews, and the training program. The objective is that continual improvement in the training program results.

The last element in the SAT process is the evaluation of the control room crews (see above example). This step is a vital step in the process and is the main area of interest, as far as this chapter is concerned. This does not mean that data collection and analysis are not important during early training or to other trades and professions, in fact they are, but these aspects are not the subjects of this chapter. The collection of data on maintenance and test operations is important and forms part of ensuring plant safety, but it is not considered here.

The idea of continuous improvement owes a lot to Deming (1992), who applied data collection and analysis to vehicle production to improve product quality and reliability. Of course, one does not treat persons as equipment, far from it. It is just the idea of exposing the crews to a variety of situations, trying to understand how they operate to determine their strengths and weaknesses, and then using this information to enhance their performance and reliability in dealing with incidents by considering the contents and methods involved in training control room crew.

To be effective in achieving the needed set of results, one should consider the performance of individuals, crews, and the training program. The context under which the crews function is the forcing function leading to both success and failure as far as performance and reliability are concerned. The meaning of context here relates to things like procedures, training, and organization. The prime reason to analyze operator responses is to understand the causes of difficulties or successes (Welsch and Sawyer, 1992). A crew may have difficulties, but unless the instructor imparts this in some form of feedback system for the training manager, then the difficulty is only seen as a local problem rather than a general issue. In the debriefing room, the crew may bring up a perceived difficulty with a procedure. The instructor may then identify this as something to train on. Later, a committee may decide if this should be incorporated into the training schedule. Different training departments may have different approaches to detecting and solving these types of issues. The data collection and analysis systems will provide the information and help in the determination of the necessary actions to be followed by providing accurate recommendations for review by training management for resolution.

13.1   DATA COLLECTION AND ANALYSIS PROCESS

The data collection and analysis process consists of a number of parts, as shown in Figure 13.1. A vital aspect of the data collection and analysis system is the

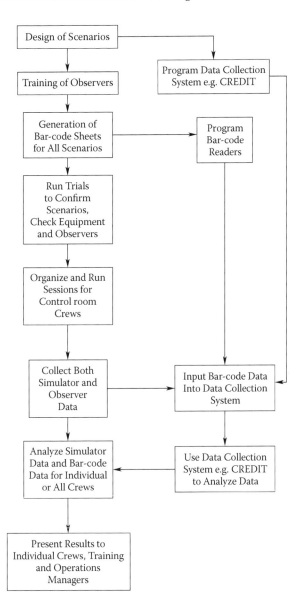

FIGURE 13.1   Data collection and analysis process.

observational taxonomy and training instructors using the taxonomy. The use of taxonomy enables two things to be accomplished: to enable a consistent representation
of the operator responses to be captured, and that this record can be processed by a
computer to draw useful conclusions about the forces driving operator responses. It
has been found from experience that it is necessary to train the instructors in the art
of making consistent observations of operator and crew performance (Montgomery
et al., 1992).

The figure indicates specific ways to collect simulator- and observer-derived data. The approach used was based on a PC connected to the simulator to collect the simulator variable data (Spurgin and Spurgin, 1994) and operator actions with a bar code reader used to record observations made by the instructors during training sessions via a bar code observational sheet (Spurgin, Bareith, and Moieni, 1996). There could be other approaches to carrying out both operations. What is important is to develop a taxonomy that forms the basis for the instructor inputs. The instructors should be trained to make consistent observations.

The collection of data should not inhibit the instructors' capability to make good observations of the interactions between the crew members, procedures, and man–machine interface with the plant (simulator). The bar code method (bar code reader and bar coded sheets) is just a way to fulfill these objectives.

13.2 ANALYSIS PROCESS

The method of collecting data has two advantages: the instructors can focus on the human-based activities, and the instructor's observations can be aligned with the corresponding responses of the plant (simulator). There are three uses of data for training and they will be discussed next.

Helping the individual crews, plots of plant variables (such as steam generator pressure) can be available for the postsession debriefing of the crews. The instructor can, during the design of the scenario, designate key actions to be checked. The list of correct actions taken and those that are missed can be printed out. The observations made by the instructor are time stamped, and these can be aligned with both the plant variables and with the actions taken by the crew. Now because of the time stamp, the instructor can quickly select the appropriate portion of the video during a debriefing. Video recordings are extensively used in training sessions, but without time stamps, the sections of video records needed by instructors are difficult to find.

The advantage of producing a bar code form is that it introduces a measure of consistency to the observations of the instructors. Additionally, because the observations are recorded and conform to a given pattern, they can be processed and sorted through for common issues and trends; for example, if the crews have a problem with part of a procedure, this can be presented to the procedures group to take some action to modify the procedure. If this aspect were missed, then the normal solution would be to increase the degree of training. Training time is a precious commodity; more time spent on one topic leads to less time on another topic. This is a net zero game. There is only a limited time available to train the crews, and any diversion can lead to wasting the training assets.

The data collection and analysis system can be used to determine inconsistencies between crews. The observational and time responses of the plant can be used. Of course, this means that identical scenarios should be run for the complete station crews. Of recent years, there has been a move to use a scenario data bank so that, for requalification tests, a random selection of test scenarios can be made. The idea of using the data bank for this purpose was based on the belief that crews would enlighten the following crews as to which scenarios would be run; thus their results would improve. It has been shown by actual tests that even if there is some information transfer, it has no effect. The advantages gained by using the data bank this way do not make up for the loss of comparative data between crew performances. The sequence of scenarios can be varied and break up the identification. One crew trying to inform another crew about the characteristics can actually handicap the other crew, especially if the sequence of accidents is changed.

A scenario consists of a sequence of human actions. One can learn a great deal plotting each individual action for each of the crews and then examining the resulting curve (TRC). Crew performance is variable and generally conforms to well-known statistical distribution. If this is the case, then the crews are effectively performing correctly. However, if one or more crews results deviate (Spurgin et al., 1992), even including failing to respond, then one should examine plant and observer data to look for reasons for the deviation. The deviations can have many causes, among them, control board instrumentation issues, procedures with poor human factors, and poor communications protocols. The strength of the data collection system is that it can be used to pinpoint the causes beyond assuming the cause is a lack of training. Having determined the likely root causes, those related to training can be considered. Examination of the strengths and weaknesses of the crews taken together can reveal topics for which additional training should be carried out and for which training can be deemphasized. The result is a training program that more accurately focuses on the real needs, rather than on the perceived needs. This can lead to safer operations.

The analysis of the data can assist training and operations managers in their review of the overall performance of the crews and the training program to determine their best strategy in enhancing the safety and availability of the plant. This outlook is obtained by looking at the total picture of crews where they are performing well and not so well, how many times problems with sections of procedures arise, and how effective the control board instrumentation is in helping the crews during accidents. One can see if overall performance of the crews, the training program, and all of the support systems are working well and getting better or worse. The question that should be asked is: Is there continuous improvement in all of these aspects? One can also relate other performance measures, like the number of plant trips, to these data and draw conclusions on the health of the plant organization as a whole. By examining the data sets over the years, one can appreciate whether or not crew performance has been improving. One can also get an idea of issues continuing to affect performance and what items have been solved and have not reappeared.

13.3   EXPERIENCE WITH DATA COLLECTION

Data collected during training simulator sessions has been used very effectively for debriefing sessions to support the interactions between the instructors, the shift supervisor, and the crew. The instructor specifies the set of plant variables and a list of operator actions to be monitored relative to the training objectives of the lesson plan. He then collects and reviews the data actually collected, including his set of observations recorded during the execution of the scenario, by using the processed data retrieved from the data collection system (Spurgin and Spurgin, 1994). If necessary, he can decide what sections of the video record to identify from the plots and spreadsheet of the observation data. With this information, he is in a good position to review the crews' responses during the debriefing session (Welsch and Sawyer, 1992).

The record of the scenario responses forms part of the permanent record that is available to other instructors or management. The individual instructor can discuss all the actions taken by the crew, so if they miss a procedure step, it is noted and discussed at the time. Also, as a result of the debriefing session and the observations, it may be possible to determine the cause, such as the procedural lack of understanding or poor human factors design, of the procedure.

However, when the record of all of the crews indicates a problem in the same area, then it is not a single crew issue, but something affecting the majority of crews. This process can focus the interest of management. If after a review, it is decided that it is a training problem, then the matter is dealt with by the training department. However, if it is considered to be a procedures matter, it is then turned over to the procedures group. When a change to the procedure is defined as necessary, the timing is determined by reference to the risk associated with not making a change. If the risk is high as determined by the PRA, then the matter should be attended to quickly. The discussion could apply to aspects other than procedures and training.

A review of data collected during accident scenarios (Spurgin et al., 1990a) indicates problem areas associated with missing steps in procedures for various reasons, using knowledge and not procedures, making the wrong decisions relative to, say, tripping reactor coolant pumps, misreading a situation, and taking actions as though they were a major leak as opposed to a small leak.

A function of the bar code sheets as part of the data collection system is to enable the instructors to designate who took an action, what the action was, what procedure was being used, and how the decision was arrived at. This kind of information cannot be determined from just the simulator input. The instructors can also define areas of interest as far as the video records are concerned. The video record can be used to support the observations made by the instructor and illustrate the situation for the crew. However, videos can be time consuming to search through to locate the exact area for the crew to review during the debriefing.

Separating the functions of the instructor into "simulator related" and "observer related" and recording these inputs can give a more complete picture of the crew response. Having taxonomy for observations and training the instructors in its use means that a consistent picture of all the crews' performance can be gained. It has been found that observers without training are not consistent in their evaluations and often have biases in their observations (Montgomery et al., 1992). Depending on the aggregated data, management can decide how to pursue excellence. Should it be necessary to understand an actual plant incident, one can consult the collected data from the simulator for a similar transient to understand how the operators act under this circumstance. The topic of human error often crops up, by this it is meant that the operators have made a mistake. By having accident data and simulator data, one can more readily understand what really caused the accident.

The data collected are useful not only for training purposes, but also for other uses. Table 13.1 lists some of the uses.

The primary use of the data collection is to support training; these other uses can help share the cost of the simulator and staff. Some of the above have had

TABLE 13.1

Uses of Simulator Data

Support human reliability studies

Support plant changes

Validate procedure changes

Show consideration of operator versus automation for protection actions

Support licensing responses to U.S. Nuclear Regulatory Commission questions

Check new operator displays and aids

Conduct simulator verification

application. Support of HRA studies has been carried out both in the United States and Hungary. In the United States, Virginia Power used the simulator for their HRA in their IPE (Virginia Power, 1991). Paks NPP carried out extensive simulator sessions over a period of years and applied the results to their PSA (Bareith, 1996). The Czech Research Institute, REZ, and the utility used the simulator for validation of the symptom-based procedures (Holy, 2000). Texas Utility used simulator studies along with the use of the emergency shutdown panels in their Fire PRA study. A research project jointly sponsored by Taiwan Power and EPRI used the simulator to compare hard-copy procedures and a computerized support system at Kuosheng BWR NPP. The process validated the procedures and indicated the superiority of the computer system as far as affects recovery from cognitive errors by the control room crews (Spurgin et al., 1990b).

The extensive use of simulator data for the Paks PRA (Bareith, 1999) and in the Surry IPE (Virginia, 1991) for HRA purposes has meant that the PRAs really reflect the NPP, not some mythical plant. The use of the simulator to validate the new symptom-based procedures for a VVER were first tested to determine technical errors and then validated at Dukovany NPP (Holy, 2000). One use for the data collection installed at Diablo Canyon was the capability of supplying data to substantiate the timely actions of the operators to prevent unnecessary use of the RHR system under certain conditions. By examining the set of computer records of the actions by the crews responding to a spurious safety injection signal with the timely termination of the RHR pump operations, PG&E was able to assure the USNRC, with data, that the RHR system would not be operating under shutoff head conditions for any length of time that might lead to RHR pump failure.

The data collection systems have been used for simulator verification at various U.S. plants. These systems have the capability to do more to track operator actions. Plant variable data displayed in various formats are used in the postscenario debriefing sessions as part of control room crew training. Human factors changes expected to have led to operator performance improvements have actually led to a reduction in the performance of the operators. Failure to carry out all steps in a procedure has been detected. For example, some 13 isolation steps are required following a steam generator tube (SGTR) accident, but it has been observed that many crews took only three steps. It was good that these were the most significant isolation steps.

Electricité de France has made extensive use of simulators for both HRA and human factors purposes over a long period of time from 1980s to recently (Le Bot et al., 2008; Villemeur et al., 1986). LeBot is associated with MERMOS for HRA purposes, and Villemeur was using the simulator sessions for investigations into the use of a workstation for engineers assisting in controlling NPP accidents. This shows that simulators can be effectively used for purposes other than just operator training. Paks also used simulators for combined training, HRA, and human factors purposes (Bareith et al., 1999). The Czechs have used simulators for EOPs validation (Holy, 2000); this is more than checking the technical validity of the EOPs. These exercises were combined with ensuring that the operator actions that were taken were correct based upon the EOPs.

13.4   CONCLUSIONS AND RECOMMENDATIONS

Experience has shown that data collection and analysis systems can reveal insights into operator responses that are difficult for instructors to fully follow and record because of the complexity of the accident responses.

It has been shown that these systems can be used effectively for a number of other activities, including human reliability studies, validation of emergency operating procedures, and determining when operators perform actions in compliance with and outside of EOPs. Thus the systems do what they are designed to do.

The use of these systems as a normal part of a training regime requires changes in the way that training departments collect and analyze data. By going to more automated data collection and analysis systems, the workload of the instructors is reduced in the area of making careful observations of operator actions during training sessions. During these sessions, the instructors are now tasked with noting the operators' use of procedures, communications, and interactions with the control instrumentation and operator's aids. The rest of the information required by the instructors can be obtained from the data collected from the plant simulator.

Data can reveal actions taken by operators that may not be compliant with procedures and training. Only the careful review of the data can reveal subtle errors. As with most control systems, strong feedback ensures that a controller corrects deviations quickly and effectively. This is also true of the training process. Strong feedback derived from the correct use and understanding of the data collected from simulated accidents can improve operator performance and reliability, leading to enhanced plant safety and availability.

It is recommended that training departments consider moving to a more data-driven feedback process to enhance the quality of response to accidents and improve plant-operating reliability. Current methods used by many training departments are to use the instructors as observers and evaluators of crew and individual performance. The process has a tendency to introduce biases into the evaluation process (Montgomery et al., 1992), especially when a crew is involved in two or more scenarios in the same session. A better way is to collect the data in each case and then base one's evaluation on what was recorded. The process ought to be split into two distinct steps: data collection, followed by a separate evaluation.

14 Discussion and Conclusions

INTRODUCTION

The discussions below cover a number of different areas from those specifically associated with HRA methods, techniques, and data connected to risk assessment. There are items covering organizational arrangements and the design of hardware. The headings below are not in any given sequence, but they do relate to the depth of the impact of HRA in the business of design and operation of plants in high-risk industries. It is pointed out that HRA is not just the selection of human error probabilities to fill a need in a PRA/PSA study. Because the reliability of humans affects both the economical and safe operation of plant, the HRA experts need a deep understanding of all aspects that affect the plant, from training to design and operation. The HRA expert needs to draw on the skills and knowledge of plant experts, including equipment designers. They also have to understand how plant management operates to understand how their decisions can potentially affect plant safety. The HRA experts should work closely with the rest of the PRA team so their views of the plant can be incorporated into the logic of the PRA, and it can reflect the real risks to plant operation.

14.1 HEP RANGE

The focus of the book is on human reliability and associated methods. One should realize that humans are capable of working reliably, but also have limitations with respect to their reliability. In other words, the reliability range of humans is limited, from say 1.0 (unreliable) to 1.0E-3 (quite reliable).

Designers and management sometimes have unreasonable expectations as to the reliability of humans and do not figure reality into their designs and corresponding costs. If there is a requirement that accidents will not occur (meaning held to a very low frequency), one needs to design the system to accomplish this. It is both difficult to design and maintain with low probability of failure rate systems relying principally on humans. To achieve high-reliability systems, one needs to consider the totality of the design by the use of barriers, safety systems, and humans, working together.

The postcommissioning state needs to be organized to maintain high reliability—this calls for equipment maintenance and test, and the corresponding things for humans, such as training and testing. These actions are needed to prevent the slow degradation of resources leading to lower overall reliability. Care is also needed to ensure upgrades and modifications are held to the same high standards.

Exactly how reliable humans are depends on the situation or context under which they are operating. This is the thesis of this book, and the method of calculating their reliability depends on the HRA method chosen.

14.2 IMPACT OF THE SELECTION OF HRA MODELS

The choice of the wrong method can incorrectly imply that humans are more reliable or less reliable in performing a given task. One cannot make a decision as to the reliability of humans without understanding their operational context. It seems that this message is not understood by many persons. One comes across people who state what they believe about human reliability without any sense of trying to understand the complexities of situations about which they are opining. This goes both ways. Sometimes it is their opinion that humans are completely unreliable; they cannot do things without making mistakes. And in other cases, they assume that humans are highly reliable. These opinions can even be present in the same organization.

14.3 ORGANIZATIONAL INFLUENCE

The other problem is that some managers seem to think that they are infallible in making decisions (gift of God?), where they are not privy to details known to others, whose advice they disregard. This is all too prevalent in technical leadership. The case history behind NASA launches illustrates this point very well, but NASA is not alone. In a case involving a nuclear power PRA study, in one scenario it was estimated that the chance of success of the control room crew detecting the presence of a failure in a remote cabinet due to fire was 50%. The presence of a fire in the cabinet was indicated on the control board by a series of lights. The lights were poorly placed and sized. The assessment was that the crews would not detect the impact of the fire. Essentially, it was a coin toss. The plant management disagreed with me and said it was at least lower than 1.0E-3. At that moment, it was not possible to prove either evaluation. However, shortly afterward, a simulator was installed and a scenario was run based on the above scenario and the crews failed.

Any safety assessment must include the concept of limitations in the reliability of humans. By this is not meant that the humans are unreliable, but their reliability bounds need to be understood. In the right circumstances, humans are very reliable. They are not ultrareliable. More like 1.0E-3 than 1.0E-6. When you need an ultrareliable system, you need a system of barriers, and controls as well as humans. You cannot just rely on a person to provide a high level of reliability. People make mistakes, so the system has to be designed to reflect what is possible.

An accident recently occurred on a train line in Los Angeles. The safety of the passengers was in the hands of the driver, and the warning given to him was a red light. A pretty usual design for railways, and generally it works. However, the underlying error probability in this situation is less than 1.0E-3, more like 1.0E-2. The problem was that the driver was distracted (text messaging). The passenger train did not stop, and it was hit by an oncoming freight train. The general conclusion is that the train driver caused the accident, which is correct. However, the possibility of a similar accident is likely due to the fallibility of humans. The design of the system was poor, and the interplay between the train companies (passenger and freight) was not conducive to operating a highly safe rail system. The basic problem is the single-line rail design without automatic traffic control and the possibility of both types of trains occupying the same track at the same time.

14.4 LESSONS FROM THE ACCIDENT REVIEWS

A number of accidents in different industries are covered in Chapter 8. The general message is that the PRA/PSA structure is quite appropriate to study the possibility of accidents. However, it is important for the PRA/HRA analysts to realize that the boundaries for the investigations have to consider not only that an operator may make errors, but that the operating conditions may also be different than those described by plant manuals and in discussions with plant management and staff.

The importance of design and management decisions has a considerable influence on the potential of an accident and its consequences. The anatomy of the accident reveals how certain decisions or sometimes the lack of decisions can lead to an accident that might have been avoided. For example, the Tenerife accident points to the lack of awareness of the airport staff impact on airport operations of suddenly introducing a number of Boeing 747s in a regional airport that normally deals with much smaller airplanes. One issue was the amount of space taken by the parked 747s. There was a failure to identify the special need to confirm that the runway was physically clear before allowing the KLM 747 to take off. Running a PRA related to this problem may not bring to the fore the altered circumstances of runway operations, because of the numbers of 747s.

The TMI Unit 2 accident brought to light differences between the views of the crew as to the transient performance of the plant versus those of the analysts. Supposedly, an analyst might assume that the crew understood the characteristics of the plant and as a result would have continued to feed the reactor with safety injection water and realized that the increase in level in the pressurizer was due to boiling in the core. Here the failure of the analyst to understand the operators' level of understanding could have led to the assumption that the reactor was safe, if tested by failures in the feed water system. The model of the NPP performance was a problem for the crew, but the failure of an analyst could lead to the pronouncement that the NPP was safe based on his assumptions of crew knowledge and understanding. This error could have led to an incorrect assessment of crew reliability with an HEP being set to 1.0E-3, instead of 1.0E-1. This further reinforces the need for the HRA analyst to be knowledgeable about plant operations as well as HRA modeling techniques.

Some of the lessons from the accidents are as follows:

1. Unreasonable expectations for the reliability of humans—see the Flaujac rail accident
2. Possible interactive effects between operations and maintenance—see Piper Alpha accident

3. Assumed operational limits dependent on instrument readings, which may fail, leading to cascading events—see Texas City accident
4. Assumption that the staff will continue to follow normal operating practices, when the staff is facing economic uncertainties—see Bhopal

14.5 SELECTION OF EXPERTS

The HRA expert needs to be a well-rounded individual knowledgeable in design and operation of plants, with an understanding of operator training and experience as well as an understanding of HRA tools and methods. In the past, the HRA task has been delegated to the least experienced person on the PRA team. Clearly this is wrong given the importance of humans in the assessment of risk. The real requirements for an HRA team member are extensive, covering many fields, including knowledge of HRA methods and data.

Review of PRAs teams indicates that most of them do not seriously tackle the issue of expertise in HRA; often the least-qualified individual is selected to "look after" HRA. This is unfortunate because the impact of humans on the risk of operation of technical plants like nuclear power, chemical, and oil and gas industries is high. Examination of accidents indicates that the decisions made by management without reference to the human side of operations almost always lead to increased risk. The whole operation of the plant from design, construction, and operation needs to be seen from the point of view of the relationship between equipment and humans. Safety is not just an equipment requirement, it involves the design of the various interfaces, how the equipment is to be maintained and tested, training of the personnel, and monitoring of the plant in such a manner that the operational personnel get early warning of difficulties.

In order to achieve the above objectives, persons with an awareness of human reliability need to be involved to advise the management team in the cost–benefit of implementations. This needs a deep understanding not only of HRA models and techniques, but also of the design alternatives and their payoff from the point of view. The plant needs to be designed from the beginning with both safety and operability in mind. Expecting the operational staff to make up for design deficiencies by training is not a good way to go. Only so much can be achieved by training, and it is a zero-sum game. The impact of training wears off and has to be reinforced. Only a certain amount of time can be spent in training. The amount of time spent in training nuclear power crews is considerable; not all industries can justify this amount of training. The time spent has to be more cost effective, so by examining the impact of human reliability on the operation, one can better appreciate how to improve plant safety and operational efficiency. This also goes for temporary modifications to an operating plant. Before changes are made, the consequences of the modification need to be investigated. The following questions need to be asked: How is it to be operated? Who is going to operate it? Under what conditions is it to be operated? Is the understanding of the operation crew up to carrying out the operation safely? What is the effect of the change on the information presented to the crew? What is the procedure if something goes wrong? As one can see, many of the questions raised can only be answered with respect to the impact on humans, because they are resources to accomplish the operation under these changed conditions, but they need also to be prepared. Examples of the degree of preparation are the maneuvers undertaken by the military before an attack on an enemy installation.

14.6 DATABASES

A significant issue in the HRA process is the availability of a validated database of human errors that can be used in the PRA/HRA process. Apart from the validity of the selected HRA model, one is faced with the questions: Is the database correct for this application? How do we know? Uncertainty with the data associated with the THERP was there from the first. The uncertainty in the HEP data was tackled by introducing uncertainty analysis into the process, but this did not really answer the question and just glossed over the issue. At the time, it was thought that the human contribution to risk was low. A figure of 15% contribution to risk was assessed, so the emphasis on human activities was lower than it should have been. The belief was that the automated, highly reliable safety systems for reactor protection were the main barriers to accident progression. Later, it became clear that the role of humans was important to preventing, mitigating, and terminating accidents. In time, the contribution increased from 15% to 60% to 80%. Therefore, it becomes more important to ensure the accuracy and validity of the human error database and how it should be used.

The issue has become sufficiently important for the use of one HRA method, HEART (Williams, 1988), to be questioned despite the fact that it has been used in the United Kingdom for the last several years for a variety of PRA studies. A method has been proposed to replace HEART, called NARA (Kirwan et al., 2005). NARA's improvements have not been limited to the database. British Energy called for steps to be taken to validate the NARA process, including the database, and a paper has been written covering the validation process by the use of HRA experts (Umbers et al., 2008). This process is a good step in the right direction to approach the issues behind the use of HRA methods.

It is not clear at this moment if NARA and its database, based upon CORE-DATA (Gibson et al., 1999), are approved by British Energy. This could be because British Energy is being acquired by Electricité de France (EDF) at this time.

More discussion on the topic of databases is presented in Chapter 11. The method advocated by the author is to draw on the knowledge of domain experts aided by the use of a set of designated influence effects and anchor values (i.e., HDT method). Also, the characteristics of the HRA choices should be confirmed by simulator experiments, where possible.

14.6.1 Fires

Fires, low-power and shut-down and fuel repository operations are mentioned because they present different situations for human error generation. Fires can break out in various locations within the plant and can affect multiple systems by virtue of power and control cables being destroyed. This presents for the operational crew significant problems. The fire can lead to an incident, so removal of the means of dealing with the accident may at the same time provide confusing information to the operators. A key part of the analysis is for the operators to determine that the situation is caused by a fire and not a combination of other effects. The fire detection system plays a key role, and after that the crew has to determine which systems are affected and take the necessary steps to stabilize the plant. A fire brigade responds to fire alarms and takes action to suppress the fire, using appropriate means. Fires can occur within control room cabinets, and this can lead to the evacuation of the control room. In this case, sufficient instrumentation and controls should be available to the crew in an alternative location.

The steps in the fire analysis process are related to detection and awareness of the fire and its location. Hopefully the crew has the information to inform them of the possible impact of the fire, because of its location, and availability of safety equipment, and also help resolve the confusion factor because of spurious alarms and indications in the control room. Part of the problem with fires is that they may cause shorts and breaks. This makes things more difficult for the crew in their diagnosis. There are also issues associated with the response timing of the fire brigade. Sometimes the function of firefighting is split between onsite and offsite groups. Modeling of the crew's responses may be difficult, but the main thing is availability of useful information to be presented to the crew in a timely manner.

14.6.2 Low-Power and Shut-Down

Low-power and shut-down operations are more complicated than full-power operations because of issues related to the accuracy of instrumentation and operation of equipment designed for full-power operation. Additionally, the functionality of the control room crew is affected by various operations to either get the plant ready to shut down or the variety of operations being carried out during shut down. During shut-down conditions, many additional tasks are undertaken, such as refueling and equipment refurbishment. These operations increase the numbers of staff on-site and disperse the control over the operations. The control room staff is now just one element out of many elements carrying out operations. This affects the supervisory function of the crew; hence, their effective reliability is reduced. Some operations are controlled by persons and are not automatically monitored and reported within the control room complex, so accidents can occur. This means that the range of accidents is different and hence the human actions now become wider than just the control room staff and its support staff. The safety system may have lower redundancy and lessened capability. All of these things increase the incremental risk of operation, somewhat balanced by two effects of lower initial power levels (reduced decay heat) and longer times to critical conditions. Unfortunately, the availability of HEP data is much reduced due to the fact that the plant simulator does not cover these conditions, and expert judgment is lessened by virtue of reduced duration of these periods. Each of the accident situations would need to be examined very carefully before selecting an HRA model or method.

14.6.3 Fuel Repository Operations

Fuel repository operations present a different situation than the usual power plant operations: the description that follows is very simplified compared with the actual handling processes. Spent fuel arrives at the facility via train or truck in utility-supplied shielded casks and is handled in a number of different ways depending on the contents of the casks. Sometimes it just moves through the process, being packaged and repackaged into different forms and then sent to the final repository. Sometimes the fuel casks are sent to a location on the site to age (i.e., left for some while so that the decay heat falls to an acceptable level). Once the spent fuel casks reach the right heat-generation rate, the casks return to one of the facilities for repackaging. If the fuel needs special handling, because of its state, it is sent to another facility and repackaged into an acceptable form. This all sounds vague, but in actuality the operations are well engineered.

As stated above, the operations are different than those for nuclear power plants. In nuclear power plants, accidents are possible but not likely. Once an accident occurs, the crews take action to mitigate or prevent the accident. Recovery is an important feature in nuclear power plant safety.

In the case of the repository, accidents can occur by dropping fuel casks, causing the casks to open and release radioactivity. The design of the casks is central to the prevention of releases. However, the casks coming from the utility sites are different than those that eventually contain the spent fuel rods in the repository. The repository is a large number of facilities whose function is to move spent fuel in one grouping to another and then transport the spent fuel within casks to the final repository. No processing goes on at the facilities, only moving, and reorganizing the spent fuel in various types of casks.

The human activities deal with handling spent fuel safely and taking precautions to prevent damage to the casks. Several steps are taken to prevent damage, such as limiting the speed of motion of transport vehicles, so that if a cask is hit it will not break. Exposure of spent fuel rods can occur, but movement of fuel is held within confinement-type rooms. Control of the fuel is done remotely, so that exposure of humans to radiation is very controlled by the use of locked entry doors and interlocks to prevent motion and control speed.

The focus of the HRA study is the study of events that lead to dropped, crushed, or damaged spent fuel casks that could lead to radiation release. Some data are available on the dropping of loads from forklifts and cranes, but other human errors have been considered, and the data for these cases have been based on a combination of the use of HRA models like NARA, HEART, THERP, and CREAM, as appropriate. Expert judgment has been used to supplement the data within these methods and to estimate the effectiveness of various performance shaping factors (PSFs) or error producing conditions (EPCs). The HRA study is quite different than that of a standard PRA/PSA for an NPP, and therefore, the methods were chosen with the actual situation in mind. Later, once the facility is functioning, data will become available to update the HEPs, EPCs, and weighting factors, assessed proportion of affect (APOA).

14.7 RELEVANCE TO OTHER INDUSTRIES

Clearly the basis for the use of HRA methods has been the nuclear power business; in fact, THERP and the handbook approach started not in the power business but were nuclear related because of the risk effect. Seemingly, the risk effect has had prime influence upon the development of PRA/PSA and HRA. Other industries have followed a different path, because the perceived risk is thought to be lower. The public has grown accustomed to the inherent risk of operation involving these different industries. The consequence of a large nuclear accident is very severe; however, the probability of it occurring is very low, so the risk is low as demonstrated in practice over these last years. However, in many of these other industries, the risk is in fact higher, but because of public and regulatory perception, the need for better protection methods is not obvious and steps for controlling risk have not been taken. Chapter 8 covers a number of accidents, and these indicate that accidents with large consequences are not confined to the nuclear industry. The utility of PRA/HRA techniques has been demonstrated. There appears to be no reason why other industries could gain relatively as much as the nuclear industry by the application of HRA techniques. One can go just so far with the application of HAZOP techniques. Ideas coming from PRAs can help in understanding that the interactive effects of failure in one piece of equipment can influence the operating conditions of other parts of a plant. The Texas City and North Sea Piper Alpha rig accidents indicate that the sequence of events is very important.

The human actions, design, and management decisions can have a dominant influence on accident probabilities. These kinds of things are best determined by the application of the discipline of PRA and the use of its associated technique of HRA. Studies indicated the importance of the human contribution to accidents. It is not all about equipment; people are involved at all stages.

As the study of accidents has pointed out to us, sometimes it is the designer many years prior to operations, the maintenance operator not returning equipment to the right condition, its manager failing fund upgrade training; it is the crew failing to correctly understand a given procedure, and so on, that cause accidents.

The impact of these things should be considered in the development of the PRA/HRA to determine what is important. For example, if the Metro-Line (*LA Times*, 2008) management had realized that there was a limit to reliability of train operators, then they should have included an automatic train stopping system in the design and worked out a better way of coordinating the passenger and freight trains operating on essentially on the same track.

14.8   RELEASE OF DATA AND MODEL DEVELOPMENTS

Some time ago, Apostolakis (a professor at MIT) made a comment that data from the EPRI ORE project was not freely available to researchers for review and comment because of constraints on publication of EPRI, and this made the ORE report much less useful than it should have been. Clearly, some things are propriety and belong to the owners, who decide what to do with them.

However, in the case of both HRA methods and data sources, it is the considered opinion of the author that this information should be released so that the HRA community can act as a community and help in the growth of techniques and sources of data. In the end, the industry and the general public will be safer, and the plants will operate more efficiently if information flows freely. Critical review of methods will lead to better methods. The parochial forces that control the data and methods should realize that they can gain more in the long term by their release.

Since the Three Mile Island Unit 2 accident and the actions taken by the USNRC and later by others, the use of simulators has increased tremendously. As part of training, sessions have been held on a regular basis duplicating difficult to severe accidents without the intervention of instructors. The mantra goes up that there is no human error data, or the data used in studies are questionable. The worldwide industry has had the opportunity for some 20 years to collect data on these simulator sessions. This would have provided a huge database, perhaps with too much information to analyze. Why has this not been done? Is it too difficult to collect, or is there some other reason? Well it is not too difficult to collect, and this has been demonstrated many times. So the answer lies elsewhere. A concern is that the information may be used against the operators and utilities. This is a legitimate concern, because we live in a blame-based society (Whittingham, 2005). Data can be collected, if the crews and operators are anonymous.

This is not really possible in the case of the utilities, unless a separate company is set up to collect and to analyze the data. The FAA's Aviation Action Safety Program (ASAP) may act as a possible model system for reporting. Unfortunately, some airlines appear to be using the system to punish pilots, against the objective of the program.

One can see the advantage of such a system. There are 105 U.S. plants for which there are six crews per plant. Each crew is exposed to two sessions per year, and each crew is likely to have to deal with five scenarios per session. In each session there are one to four human events of interest (maybe more). This means that for each year, there should be recorded data on 2520 to 25,200 human events per year just for the United States, so what about the rest of the world? Somebody said they were short of HRA data, as well as statistical data and the answer is probably yes, but we should not be short of useful information about how NPP crews respond to accidents. We should be able to define what events we need to consider and how changes in HSI and EOPs affect the responses. We also should have had data from 1985 to the present time.

Using HEART, NARA, and CREAM, one is limited to a small number of tasks, the generic task types (GTTs), numbers of which are, respectively, 9, 7, and 13. The emphasis in all of these models was on error producing conditions (EPCs), and these were, respectively, 38 and 10 for HEART and NARA. For CREAM, Hollnagel used the term common performance condition (CPC), and his grouping is nine main effects and from two to four level breakdowns. For the HDT, the number of influence factors (IFs) can be varied, but about six are normally selected. For each IF, the normal number selected is from three to four possible levels. The level is determined by the impact of a specific accident on the particular IF, so one goes back to collecting useful data from simulators to help define the importance of IFs and quality factors (QFs) or their equivalents.

References

Andresen, G. et al., 2003, "Experiments in the Halden Human Machine Laboratory (HAMMLAB) Investigating Human Reliability Issues," Towards Conference of Technical Nuclear Safety Practices in Europe, EUROSAFE, Paris, France.

Anonymous, 2008, "Tenerife disaster," report, Wikipedia, date unknown, latest modification added in May 2008.

ANS, 2005, "Decay Heat Powers in Light Water Reactors," American Nuclear Society, La Grange Park, IL.

AANSI/ANS-58.8, 1994, "Time Response Design Criteria for Safety-Related Operator Actions," American Nuclear Society, La Grange Park, Illinois [reaffirmed in 2001 and 2008].

NS N-660, 1977, "Draft Proposed American National Standard, Criteria for Safety Related Operator Actions." American Nuclear Society, La Grange Park, IL.

ANS, "ANS Standard on Low Power and Shut Down PRA Methodology," American Nuclear Society, La Grange Park, IL; in progress.

ASME, 2002, "Probabilistic Risk Assessment for Nuclear Power Plant Applications (PRA)," RA-S – 2002, American Society of Mechanical Engineers, www.asme.org

Bareith, A., 1996, "Simulator Aided Developments for Human Reliability Analysis in the Probabilistic Safety Assessment of the Paks Nuclear Power Plant," VEIKI Report # 20.11-217/1, Budapest, Hungary.

Bareith, A. et al., 1999, "Human Reliability Analysis and Human Factors Evaluation in Support of Safety Assessment and Improvement at the Paks NPP," Fourth International Exchange Forum: Safety Analysis of NPPs of the VVER and RBMK Type, October, Obinsk, Russian Federation.

Barnes, V. et al., 2001, "The Human Performance Evaluation Process: A Resource for Reviewing the Identification and Resolution of Human Performance Problems," NUREG/CR-6751, U.S. Nuclear Regulatory Commission, Washington, DC.

Beare, A.N. et al., 1983, "A Simulator-Based Study of Human Errors in Nuclear Power Plant Control Room Tasks," NUREG/CR-3309, U.S. Nuclear Regulatory Commission, Washington, DC.

Beare, A.N. et al., 1990, "An Approach to Estimating the Probability of Failures in Detection, Diagnosis, and Decision Making Phase of Procedure-Guided Human Interactions," Draft report for EPRI RP-2847_1, Electric Power Research Institute, Palo Alto, CA (Final version contained within EPRI TR-100259, Parry et al., 1992).

Broadribb, M.P., 2006, "BP Amoco Texas City Incident," American Institute of Chemical Engineers, Loss Prevention Symposium/Annual CCPS Conference, Orlando, FL.

Carnino, Annick, Nicolet, Jean-Louis, and Wanner, Jean-Claude, 1990, "Man and Risks Technological and Human Risk Prevention," Marcel Dekker, New York and Basel.

Comer, M.K., E.J. Kozinsky, J.S. Eckel, and D.P. Miller, 1983, "Human Reliability Data Bank for Nuclear Power Plant Operations," NUREG/CR-2744, Vols 1 & 2, Sandia National Laboratories, Albuquerque, NM.

Comer, M.K. et al., 1984, "General Human Reliability Estimates Using Expert Judgment," NUREG/CR-3688, U.S. Nuclear Regulatory Commission, Washington, DC.

Cooper, S.E. et al., 1996, "A Technique for Human Error Analysis (ATHEANA)," NUREG/CR-6350, U.S. Nuclear Regulatory Commission, Washington, DC.

Cullen, W.D., 1990, "The Public Inquiry into the Piper Alpha Disaster," Vols 1 & 2, HMSO, London, England (presented to Secretary of State for Energy, 19-10-1990 and reprinted in 1991 for general distribution).

Dekker, Sidney W.A., 2005, "Ten Questions about Human Error," Lawrence Erlbaum, Mahwah, NJ.

Dekker, Sidney W.A., 2006, "The Field Guide to Understanding Human Error," Ashgate, Aldershot, Hampshire, England.

Deming, W.E., 1992, "Out of Crisis," Published by Massachusetts Institute of Technology, Center for Advanced Engineering Study, MA.

Dougherty, E.M., and J.R. Fragola, 1988, "Human Reliability Analysis: A Systems Engineering Approach with Nuclear Power Plant Applications," John Wiley & Sons, NY.

Dougherty, E.M., 1990, "Human Reliability Analysis—Where Shouldst Thou Turn?" Reliability Engineering and System Safety, 29(3), 283–299.

Embrey, D.E. et al., 1984, "SLIM-MAUD, An Approach to Assessing Human Error Probabilities Using Structured Expert Judgment," NUREG/CR-3518, U.S. Nuclear Regulatory Commission, Washington, DC.

EPRI, 2005, "EPRI/NRC Fire PRA Methodology for Nuclear Power Facilities (NUREG/CR-6850)," Electric Power Research Institute, Palo Alto, CA.

Fennell, D., 1988, "Investigation into the King's Cross Underground Fire," The Stationary Office Books, London.

Feynman, R.P., 1986, "Appendix F—Personal Observations on the Reliability of the Shuttle," Part of the Rogers' President Commission Report on the Space Shuttle Challenger Accident.

Feynman, Richard P. with Ralph Leighton, 1989, "What Do You Care What Other People Think?" Bantam Books, NY.

Fleming, K.N. et al., 1975, "HTGR Accident Investigation and Progression Analysis (AIPA) Status Report (vol. 2)," General Atomics, San Diego, CA.

Forster, J. et al., 2004, "Expert Elicitation Approach for Performing ATHEANA Quantification," Reliability Engineering and System Safety, 83, 207–220.

Forster, J., A. Kolacskowski, and E. Lois, 2006, "Evaluation of Human Analysis Methods against Good Practice," NUREG-1842, U.S. Nuclear Regulatory Commission, Washington, DC.

Forster, J. et al., 2007, "ATHEANA User's Guide, Final report," NUREG-1880, U.S. Nuclear Regulatory Commission, Washington, DC.

Frank, M.V., 2008, "Choosing Safety: A Guide to Using Probabilistic Risk Assessment and Decision Analysis in Complex, High Consequence Systems," RFF Press, Washington, DC.

Gertman, D.I. et al., 1988, "Nuclear Computerized Library for Assessing Reactor Reliability, NUCLARR," NUREG/CR-4639, Idaho National Engineering Laboratory, ID.

Gertman, D. et al., 2004, "The SPAR-H Human Reliability Analysis Method," NUREG/CR-6883, Prepared for the U.S. Nuclear Regulatory Commission, Washington, DC.

Gibson, W.H. et al., 1999, "Development of the CORE-DATA Database," Safety and Reliability Journal, Safety and Reliability Society, Manchester, UK.

Hall, R.E. et al., 1982, "Post Event Human Decision Errors: Operator Action Tree/Time Reliability Correlation," NUREG/CR-3010, U.S. Nuclear Regulatory Commission, Washington, DC.

Hallbert, B. et al., 2007, "Human Event Repository and Analysis (HERA): Overview," Vol. 1 and "The HERA Coding Manual and Quality," Vol. 2, NUREG/CR-6903, U.S. Nuclear Energy Commission, Washington, DC.

Hannaman, G.W., and A.J. Spurgin, 1984a, "Systematic Human Action Reliability Procedure (SHARP), EPRI NP-3583, Electric Power Research Institute, Palo Alto, CA.

Hannaman, G.W., A.J. Spurgin, and Y. Lukic, 1984b, "Human Cognitive Reliability Model for PRA Analysis," NUS-4531, Draft EPRI document, Electric Power Research Institute, Palo Alto, CA.

Higgins, J.C., J.M. O'Hara, and P. Almeida, 2002, "Improved Control Room Design and Operations Based on Human Factor Analyses or How Much Human Factors Upgrade Is Enough?" IEEE, 7th Human Factors Meeting, Scottsdale, AZ.

Hollnagel, E., 1993, "Human Reliability Analysis: Context and Control," Academic Press, NY.

Hollnagel, E., and P. Marsden, 1996, "Further Development of the Phenotype-Genotype Classification Scheme for the Analysis of Human Erroneous Actions," European Commission Publication EUR 16463 EN, Joint Research Center, ISPRA, Italy.

Hollnagel, E., 1998, "Cognitive Reliability and Error Analysis Method: CREAM," Elsevier Science, Amsterdam.

Hunns, D.M., 1982, "The Method of Paired Comparisons," in High Risk Safety Technology, A.E. Green (Ed.), John Wiley, Chichester, UK.

Holy, J., 2000, "NPP Dukovany Simulator Data Collection Project," Czech paper on procedures at PSAM 5, Osaka, Japan.

IEEE, 1979, "First IEEE Conference on Human Factors in Nuclear Power Plants," Myrtle Beach, NC.

IEEE, 1981, "Second IEEE Conference on Human Factors in Nuclear Power Plants," Myrtle Beach, NC.

IEEE, 1997, "Guide for Incorporating Human Action Reliability Analysis for Nuclear Power Generating Stations," IEEE Standard 1082, IEEE 445 Hoes Lane, Piscataway, NJ.

IEEE, 2009, "Human Factors Guide for Applications of Computerized Operating Procedure Systems at Nuclear Power Generating Stations and other Nuclear Facilities," Draft IEEE Standard 1786 currently being worked on, IEEE 445 Hoes Lane, Piscataway, NJ.

Julius, J.A., J. Grobbelaar, and F. Rahn, 2005, "EPRI HRA Calculator™—Version 3, ANS Topical Conference on PRA, September, 2005, San Francisco, CA.

Julius, J.A. et al., 2008, "Overall Design of the International Empirical HRA Methodology Study," American Nuclear Society Conference on Probabilistic Safety Assessment 2008, Knoxville, TN.

Kalelkar, Ashok S., 1988, "Investigation of Large-Magnitude Incidents: Bhopal as a Case Study," Institution of Chemical Engineers Conference on Preventing Major Accidents, London, England.

Kemeny, J.G., 1979, "The Report to the President on the Three Mile Island Accident," originally published October 30, 1979. [There is a copy of the report at www.pddoc.com/tmi2/kemeny/index.htm]

Kirwan, B. et al., 2005, "Nuclear Action Reliability Assessment (NARA): A Data-Based HRA Tool," Safety and Reliability, 25(2), published by Safety and Reliability Society, UK.

Kirwan, B. et al., 2008, "Quantifying the Unimaginable—Case for Human Performance Limiting Values," Probabilistic Safety and Management, PSAM 9, Conference, Hong Kong, China.

Kirwan, B. et al., 2008, "Nuclear Action Reliability Assessment (NARA): A Data-Based HRA Tool," Probabilistic Safety and Management, PSAM 9, Conference, Hong Kong, China.

Kirwan, B., and L.K. Ainsworth, 1992, "A Guide to Task Analysis," Taylor & Francis, London.

Kohlhepp, K.D., 2005, "Evaluation of the Use of Engineering Judgments Applied to Analytical Human Reliability Analysis (HRA) Methods," M.S. Thesis, Texas A & M University, TX.

Kozinsky, E.J. et al., 1983, "Criteria for Safety-Related Operator Actions (SROA): Final Report," NUREG/CR-3515, U.S. Nuclear Regulatory Commission, Washington, DC.

LA Times, 2008, "NASA Readies a Shuttle for Rescues, Just in Case," Associated Press, 9/20/2008.

LeBot, P. et al., 1999, "MERMOS: A Second Generation HRA Method; What It Does and Doesn't Do," Proceedings of the International Topical Meeting on Probabilistic Safety Assessment (PSA '99), Washington, DC.

LeBot, P., H. Pesme, and P. Meyer, 2008, "Collecting Data for Mermos Using a Simulator," Ninth Probabilistic Safety and Management Conference, Hong Kong, China.

LeBot, P., 2008b, "Private Communication on MERMOS." November 24.

Leveson, Nancy, 2005, "A New Accident Model for Engineering Safer Systems," MIT series, MA.

Libmann, Jacques, 1996, "Elements of Nuclear Safety," Les Edition de Physique, France.

Lyons, M., M. Woloshynowych, S. Adams, and C. Vincent, 2005, "Error Reduction in Medicine," Imperial College, London, and National Patient Safety Agency for "The Nuffield Trust," Final Report.

Martel, Bernard, 2000, "Chemical Risk Analysis—A Practical Handbook," Kogan Page Science, London, England.

Matahri, N., and G. Baumont, 2008, "Application of Recuperare to the Safety of the French Nuclear Power Plant Industry," Draft report prepared for French Nuclear Safety Organization (IRSN), Paris, France.

Mil. Standard, 2003, "General Human Engineering Design Criteria for Military Systems, Subsystems, Equipment and Facilities," Mil Standard, 1492F, 2003, U.S. Department of Defense, Washington, DC.

Moeini, P., A.J. Spurgin, and A. Singh, 1994, "Advances in Human Reliability Analysis Methodology," Parts 1 and 2, Reliability Engineering and System Safety, Volume 44, Elsevier, Amsterdam.

Montgomery, J.C. et al., 1992, "Team Skills Evaluation Criteria for Nuclear Power Plant Control Room Crews," Draft NRC report, U.S. Nuclear Regulatory Commission, Washington, DC (Available from NRC Document Room, Washington).

Norman, D.A., 1981, "Categorization of Action Steps," Psychological Review, 88, 1–15.

Norman, D.A., 1988, "The Design of Everyday Things," Basic Books, NY.

O'Hara, J.M. et al., 2002, "Human–System Interface Design, Review Guidelines," (NUREG-0700), Rev 2, U.S. Nuclear Regulatory Commission, Washington, DC.

O'Hara, J.M. et al., 2008, "Human Factors Considerations with Respect to Emerging Technology in Nuclear Power Plants (NUREG/CR-6947)," U.S. Nuclear Regulatory Commission, Washington, DC.

Orvis, D.D., and A.J. Spurgin et al.,1990, "Operator Reliability Experiments at the Maanshan Nuclear Power Plant," NP-6951L, RP 2847-3, Joint Project EPRI and Taiwan Power Company, Palo Alto, CA and Taipei, Taiwan.

Orvis, D.D., and A.J. Spurgin, 1996, "Research in Computerized Emergency Operating Procedures Systems to Enhance Reliability of NPP Operating Crew," APG Report # 35, Prepared for U.S. Nuclear Regulatory Commission, Washington, DC.

Parry, G.W. et al., 1992, "An Approach to the Analysis of Operator Actions in Probabilistic Risk Assessment," EPRI TR-100259, Electric Power Research Institute, Palo Alto, CA.

Pesme, H., P. LeBot, and P. Meyer, 2007, "A Practical Approach of the MERMOS Method, Little Stories to Explain Human Reliability Assessment," IEEE/HPRCT Conference, Monterey, CA.

Petkov, G., P. Antao, and C. Guedes Soares, 2001, "Context Quantification of Individual Performance in Accidents," Proceedings of ESREL 2001, Vol. 3, Turin, Italy.

Petkov, G. et al., 2004, "Safety Investigations of Team Performance in Accidents," Journal of Hazardous Materials, 3, 97–104.

Potash, L.M. et al., 1981, "Experience in Integrating the Operator Contributions in the PRA of Actual Operating Plants," ANS/ENS Topical Meeting on Probabilistic Risk Assessment, Port Chester, NY.

Rasmussen, J., 1979, "On the Structure of Knowledge—A Morphology of Mental Models in a Man-Machine Context," RISØM-2192, RISØ National Laboratory, Roskilde, Denmark.

Rasmussen, J., 1986, "Information Processing and Human–Machine Interaction: An Approach to Cognitive Engineering," North Holland, Amsterdam.

Rausand, M., 2004, "HAZOP—Hazard and Operability Study," System Reliability Theory, Wiley, NY.

Reason, J.T., 1987, "Generic Error-Modeling System (GEMS): A Cognitive Framework for Locating Human Error Forms," In J. Rasmussen, K. Duncan, and J. Leplat (Eds.), New Technology and Human Error, John Wiley, London.

Reason, J.T., 1990, "Human Error," Cambridge University Press, Cambridge, UK.

Reer, B., 1998, "Conclusions from Occurrences by Descriptions of Accidents (CODA)," Questionnaire response, Task 97-2 (Errors of Commission), PWG5, NEA, OECD, Paris, France.

Richards, R., 2009, HERA, private communication from Robert Richards, Principal Investigator, Idaho National Laboratory, ID.

Rogers' Commission, 1986, "Report to the President by the President's Commission on the Space Shuttle Challenger Accident," June 6, Washington, DC.

Rohrlich, Ted, 2008, "Metro-Link Train Crash," *LA Times* The Envelop, December 12.

Rook, L.W., 1962, "Reduction of Human Error in Industrial Production," SCTM 93-62 (14), Albuquerque, NM.

Ruan, Da et al., 2005, "Intelligent Data Mining; Techniques and Applications," Part III, "Advanced Simulator Data Mining for Operators Performance," Springer pp. 487–514.

Saaty, T.L., 1980, "The Analytic Hierarchy Process," McGraw-Hill, NY.

SECY-01-0067, 2001, "Report on Support to the American Nuclear Society for the Development of Standard on Probabilistic Risk Assessment for Low Power and Shutdown," U.S. Nuclear Regulatory Commission, Washington, DC.

Spurgin, A.J., P. Moieni, and G.W. Parry, 1989, "A Human Reliability Analysis Approach Using Measurements for Individual Plant Examination," EPRI NP-6560-L, Electric Power Research Institute, Palo Alto, CA.

Spurgin, A.J. et al., 1990a, "Operator Reliability Experiments Using Power Plant Simulators," NP-6937, Volumes 1, 2, and 3, Electric Power Research Institute, Palo Alto, CA.

Spurgin, A.J., D.D. Orvis et al., 1990b, "The BWR Emergency Operating Procedures Tracking System (EOPTS), Evaluation by Control Room Operating Crews," EPRI NP-6846, Electric Power Research Institute, Palo Alto, CA.

Spurgin, A.J. et al., 1992, "Operator Reliability Assessment System, OPERAS," EPRI RP-3082-1, Vol. 1, Technical description, Vol. 2 User's manual, Vol. 3 Interface design, Electric Power Research Institute, Palo Alto, CA.

Spurgin, A.J., J. Wachtel, and P. Moeini, 1993, "The State of Practice of Computerized Operating Procedures in Commercial Nuclear Power Industry," Proceedings of the Human Factors and Ergonomics Society, 37th Annual Meeting, pp 1014–1018.

Spurgin, A.J., 1994a, "Developments in the Use of Simulators for Human Reliability and Human Factors Purposes," IAEA Technical Committee Meeting on Advances in Reliability Analysis and Probabilistic Safety Assessment, Szentendre, Hungary.

Spurgin, A.J., and J.P. Spurgin, 1994b, "On the Design of an Improved System for Data Collection and Analysis Attached to Full Scope Plant Simulator," SCS Simulation Multi-conference, San Diego, CA.

Spurgin, A.J., and J.P. Spurgin, 1994c, "A Data Collection Analysis System for Use with a Plant Simulator," Institute of Mechanical Engineers Seminar, "Achieving Efficiency through Personnel Training—The Nuclear and Safety Regulated Industries."

Spurgin, A.J., A. Bareith, and P. Moieni, 1996, "Computerized Safety Improvement System for Nuclear Power Operator Training," Joint SCIENTECH and VEIKI Report for Brookhaven Laboratory, NY.

Spurgin, A.J., 1999, "Human Reliability: Dependency Aspects," Report for PNNL, USDOE Funded for Paks NPP Safety Enhancement Program, Washington, DC.

Spurgin, A.J., 1999, "Developments in the Decision Tree Methodology," PSA '99, International Topic Meeting on Probabilistic Safety Assessment, Risk-Informed Performance-Based Regulation in the New Millennium, Washington, DC.

Spurgin, A.J., 2000, "Experience with the Decision Tree Method for Several Applications," PSAM 5 Conference in Osaka, Japan.

Spurgin, A.J., and J.P. Spurgin, 2000, "CREDIT Vr 3.1, Core, Description and Operating Manual," Report for Arizona Public Service, Palo Verde, AZ.

Spurgin, A.J., and B.O.Y. Lydell, 2002, "Critique of Current HRA Methods," Proceedings of the 2002 IEEE 7th Conference of Human Factors and Power Plants, September, Scottsdale, AZ.

Spurgin, A.J., and M.V. Frank, 2004a, "Developments in HRA Technology from Nuclear to Aerospace," PSAM 7 Conference, Berlin.

Spurgin, A.J., 2004b, "Human/System Reliability in a Changing Environment," IEEE SMC Conference, The Hague, Netherlands.

Spurgin, A.J., and G. Petkov, 2005, "Advanced Simulator Data Mining for Operators' Performance Assessment," chapter in book "Intelligent Data Mining: Techniques and Applications," pp. 543–572. Physica Verlag, Heidelberg.

Spurgin, A.J. et al., 2006, "Analysis of Simulator Data for HRA Purposes," PSAM 8, New Orleans, LA.

Straeter, O., 1997, "Buerteilung der menschlichen zuverlaessigkeit auf der basis von betriebserfahung," Dissertation an der Techischen Universitat, Munchen, GRS – 138, GRS Kohn, Germany.

Swain, A.D., and H.E. Guttman, 1980, "Draft Version Handbook of Human Reliability Analysis with Emphasis on Nuclear Power Plant Applications," U.S. Nuclear Regulatory Commission, Washington, DC.

Swain, A.D., and H.E. Guttman, 1983, "Handbook of Human Reliability Analysis with Emphasis on Nuclear Power Plant Applications," NUREG/CR-1278, U.S. Nuclear Regulatory Commission, Washington, DC.

Swain, A.D., 1987, "Accident Sequence Evaluation Program: Human Reliability Analysis Procedure," NUREG/CR-4772, U.S. Nuclear Regulatory Commission, Washington, DC.

Texas Utilities, "Fire External Event—Human Reliability Assessment," Comanche Peak Steam Electric Station, Individual Plant Examination of External Events Report ER-EA-007, Texas Utilities, Dallas, TX.

Umbers, Ian et al., 2008, "Peer Review of the NARA HRA Technique," Probabilistic Safety and Management, PSAM 9, Conference, Hong Kong, China.

USNRC, 1993, "Shutdown and Low-Power Operation at Commercial Nuclear Power Plants in the United States," (NUREG_1449), U.S. Nuclear Regulatory Commission, Washington, DC.

Vicente, Kim, 2006, "The Human Factor," Taylor and Francis, London.

Villemeur, A. et al., 1986, "A Simulator-Based Evaluation of Operators' Behavior by Electricité de France," International Topical Meeting on Advances in Human Factors in Nuclear Power Systems, Knoxville, TN.

Virginia Power, 1991, "Surry Nuclear Power Station Units 1 and 2: Individual Plant Examination," Serial 91–134A, Innsbrook, VA [HRA is appendix D].

Wakefield, D.J. et al., 1992, "Systematic Human Action Reliability Procedure (SHARP) Enhanced Project—SHARP 1 Methodology Report," EPRI RP 3206-01, Electric Power Research Institute, Palo Alto, CA.

Walters, James M., and Robert L. Sumwalt III, 2000, "Aircraft Accident Analysis: Final Report," McGraw-Hill, NY.

WASH 1400, 1975, "Reactor Safety Study—An Assessment of Accident Risks in U.S. Commercial Nuclear Power Plants," NUREG–75/014, U.S. Nuclear Regulatory Commission, Washington, DC.

Welsch, J., and L. Sawyer, 1992, "Automatic Monitoring of Control Room Crews during Simulator Sessions for Training Improvement," Proceedings of the 1992 Simulation Multi-conference, Simulators IX, Simulation Series, Volume 24, Number 4, Society for Computer Simulation, Orlando, FL.

Whittingham, R.B., 2005, "The Blame Machine, Why Human Error Causes Accidents," Elsevier, London.

Williams, J.C., 1988, "A Data-Based Method for Assessing and Reducing Human Error to Improve Operational Performance," Proceedings of IEEE Fourth Conference on Human Factors in Power Plants, pp. 436–450, Monterey, CA.

Yucca Mountain, 2008, "Yucca Mountain Repository, License Application for Construction Authority," Report for U.S. Department of Energy, Bechtel and Science Applications International Company (SAIC), Las Vegas, NV.

Appendix

LIST OF HUMAN RELIABILITY DATABASE REFERENCES

Below are a number of references to human reliability databases from the past. They have not all been reviewed, but they indicate that some work has been going on in the field of database collections as far as human reliability is concerned. These references were obtained from www.emeraldinsight.com. Some of the references are repeated in the References section of this book.

Advisory Committee on the Safety of Nuclear Installations, Study Group on Human Factors (1991), *Human Reliability Assessment—A Critical Overview*, 2, Health and Safety Commission, HMSO, London.

Askren, W.B., Regulinski, T.L. (1969), "Quantifying human performance for reliability analysis of systems," *Human Factors*, Vol. 11 No. 4, pp. 393–396.

Blanchard, R.E., Mitchell, M.B., Smith, R.L. (1966), *Likelihood of Accomplishment Scale for a Sample of Man-Machine Activities*, Dunlap and Associates, Santa Monica, CA.

Coburn, R. (1971), "A human performance data bank for command control," *Proceedings of the U.S. Navy Human Reliability Workshop*, February.

Dolby, A.J. (1990), "A comparison of operator response times predicted by the HCR model with those obtained from simulators," *International Journal of Quality and Reliability Management*, Vol. 7 No. 5, pp. 19–26.

Embrey, D.E., Humphreys, P.C., Rosa, E.A., Rea, K. (1984), *SLIM-MAUD: An Approach to Assessing Human Error Probabilities Using Structured Expert Judgment*, NUREG/CR-3518, Department of Nuclear Energy, Brookhaven National Laboratory, Upton, New York, for Office of Nuclear Regulatory Research, U.S. Nuclear Regulatory Commission, Washington, DC.

Fleishman, E.A., Quaintance, M.K. (1984), *Taxonomies of Human Performance: The Description of Human Tasks*, Academic Press, London.

Gertman, D.I., Gilmore, W.E., Galtean, W.J., Groh, M.J., Gentillon, C.D., Gilbert, B.G. (1988), *Nuclear Computerized Library for Assessing Reactor Reliability*, NUCLARR Volume 1: Summary Description, NUREG/CR-4639, Idaho National Engineering Laboratory, ID.

Hannaman, G.W., Spurgin, A.J., Lukic, Y.D. (1984), *Human Cognitive Reliability Model for PRA Analysis*, Electric Power and Research Institute, Palo Alto, CA, draft report NUS-4531, EPRI Project RP2170-3.

Hornyak, S.J. (1967), "Effectiveness of display subsystems measurement and prediction techniques," RADC report, TR-67-292.

Hunns, D.M., and Daniels, B.K. (1981), "Paired comparisons and estimates of failure likelihood," *Design Studies*, Vol. 2, pp. 9–19.

Irwin, I.A., Levitz, J.J., Freed, A.M. (1964), "Human reliability in the performance of maintenance," Aerojet, General Corporation, Sacramento, CA, report LRP 317/TDR-63-218.

Kantowitz, B.H., Fujita, Y. (1990), "Cognitive theory, identifiability and human reliability analysis (HRA)," in Apostolakis, G.E., Mancini, G., van Otterloo, R.W., Farmer, F.R. (Eds), *Reliability Engineering and System Safety: Special Issue on Human Reliability Analysis*, Elsevier Applied Science, Vol. 29 No. 3.

Kirwan, B. (1982), "An evaluation and comparison of three subjective human reliability quantification techniques," MSc dissertation, University of Birmingham, Department of Production Engineering.

Kirwan, B., Embrey, D.E., Rea, K. (1988), *Human Reliability Assessors Guide*, SRD Report No. RTS 88/95Q, United Kingdom Atomic Energy Authority (UKAEA), Warrington.

Kirwan, B., Smith, A. (1989), "Human error data collection—some problems, some solutions," IBC Conference on Human Reliability in Nuclear Power, Cafe Royal, London.

Kirwan, B. (1990), "A resources flexible approach to human reliability assessment for PRA," in Proceedings of the Safety (Eds), Reliability Symposium, Altrincham, London.

Kirwan, B., Martin, B., Ryecraft, H., Smith, A. (1990), "Human error data collection and data generation," *Journal of Quality and Reliability Management*, Vol. 7 No. 4, pp. 34–66.

Laughery, K.R. (1984), "Computer modeling of human performance on micro-computers," *Proceedings of the Human Factors Society Meeting*, Human Factor Society, Santa Monica, CA, pp. 884–888.

Meister, D. (1984), "Human reliability," in Muckler, F.A. (Ed), *Human Factors Society Review: 1984*, Human Factors Society, Santa Monica, CA.

Metwally, A.M., Sabri, Z.A., Adams, S.K., Husseiny, A.A. (1982), "A data bank for human related events in nuclear power plants," Proceedings of the 26th Annual Meeting of the Human Factors Society, Human Factors Society, New York, NY.

Munger, S.J., Smith, R.W., Payne, D. (1963), "AIR—an index of electronic equipment operability data store," *AIR*, American Institute for Research, C43-1/62-RP (1).

Paradies, M., Busch, D. (1988), "Root cause analysis at Savannah River plant," Monterey, CA, paper presented at the Institute of Electrical and Electronics Engineers (IEEE) Conference on Human Factors in Nuclear Power.

Rasmussen, J. (1981), *Human Errors: A Taxonomy for Describing Human Malfunction in Industrial Installations*, RISO-M-2304, RISO National Laboratory, Roskilde, Denmark.

Reason, J.T. (1990), *Human Error*, Cambridge University Press, Cambridge.

Siegal, A.I., Wolf, J.T., Laufman, M.R. (1974), "A model for predicting integrated man-machine system reliability," in Wayne, P.A. (Ed), *Applied Psychology Services*, Technical Report M24-72-C-12777, November.

Swain, A.D. (1967), "Some limitations in using the simple multiplicative model in human behavior quantification," in Askren, W.D. (Ed), *Symposium on Reliability of Human Performance in Work*, ARML Report, ARML-TR-67-88, May, pp. 17–31.

Swain, A.D., Guttmann, H.E. (1983), *Handbook of Human Reliability Analysis with Emphasis on Nuclear Power Plant Applications*, NUREG/CR-1273, U.S. Nuclear Regulatory Commission (USNRC), Sandia National Laboratories.

Williams, J.C., Wiley, J. (1985), "Quantification of human error in maintenance for process plant probabilistic risk assessment," in Proceedings of the Assessment, Control of Major Hazards, Topmiller, D.A., Eskel, J, and Kozinsky, E.J. (Eds), The Institution of Chemical Engineers, Symposium Series No. 93, EFCE No. 322, pp. 353–365, England.

Williams, J.C. (1986), "HEART—a proposed method for assessing and reducing human error," Ninth Advances in Reliability Technology Symposium, Birmingham, IMICHE, London, UK.

Index